Digital and Kalman filtering

An introduction to discrete-time filtering and optimum linear estimation

S. M. Bozic

Department of Electronic and Electrical Engineering
University of Birmingham

Edward Arnold

First published in 1979 by
Edward Arnold (Publishers) Ltd
41 Bedford Square, London WC1B 3DQ

Reprinted 1981, 1984

British Library Cataloguing in Publication Data

Bozic, Svetozar Mile
Digital and Kalman filtering.
1. Digital filters (Mathematics) 2. Kalman filtering
I. Title
621.3815'32 TK7872.F5

ISBN 0 7131 3410 0

Photo typeset in India by
The Macmillan Co of India Ltd., Bangalore 1

Printed and bound in Great Britain
at The Pitman Press, Bath

Contents

Preface vii

Part 1 Digital filtering 1

Introduction 1

1 Introduction to discrete time filtering 3

1.0 Introduction 3
1.1 Continuous and discrete-time analysis 3
1.2 Discrete-time processing in radar tracking 6
1.3 z-transform 9
1.4 z-transform relation to the Laplace transform 11
1.5 Problems 14

2 Digital filter characterization 16

2.0 Introduction 16
2.1 Digital transfer function 16
2.2 The inverse transformation 19
2.3 Frequency response 21
2.4 Digital filter realization schemes 22
2.5 The classification of digital filters 27
2.6 Problems 28

3 Nonrecursive filter design 33

3.0 Introduction 33
3.1 Nonrecursive filter properties 34
3.2 Design procedure 36
3.3 Design modification using windows 40
3.4 Problems 43

4 Recursive filter design 45

4.0 Introduction 45
4.1 The impulse invariance method 46
4.2 The bilinear z-transform method 51
4.3 Frequency transformations of lowpass filters 56
4.4 Accuracy considerations in digital filter design 62
4.5 Problems 63

5 Further concepts in discrete-time filtering 66

5.0 Introduction 66
5.1 Derivation of discrete Fourier series (DFS) 66
5.2 Finite time sequences – discrete Fourier transform (DFT) 70
5.3 Inverse filter 73
5.4 Optimum finite inverse filter 75
5.5 Problems 77

Part 2 Optimum (Wiener and Kalman) linear estimation 81

Introduction 81

6 Digital filtering of noisy data 83

6.0 Introduction 83
6.1 Brief review of digital filtering 83
6.2 Nonrecursive estimator 87
6.3 Recursive estimator 89

7 Optimum estimation of scalar signals 92

7.0 Introduction 92
7.1 Optimum nonrecursive estimator (scalar Wiener filter) 92
7.2 Recursive estimator from the optimum nonrecursive estimator 97
7.3 Optimum recursive estimator (scalar Kalman filter) 100
7.4 Optimum recursive predictor (scalar Kalman predictor) 105

8 Optimum estimation of vector signals 109

8.0 Introduction 109
8.1 Signal and data vectors 109
8.2 Vector problem formulation 113
8.3 Vector Kalman filter 115
8.4 Vector Kalman predictor 119
8.5 Vector Wiener filter 120

9 Examples 125

9.0 Introduction 125
9.1 Scalar Wiener filter 125
9.2 Scalar Kalman filter 126
9.3 Vector Kalman filter 127
9.4 Kalman filter application to falling body 130
9.5 Wiener filter application to falling body 133
9.6 Kalman filter formulation for radar tracking 136

Solutions to Problems 141

Appendix 144

References 154

Index 156

Preface

The availability of digital computers has stimulated the use of digital signal processing (or time series analysis) in many diverse fields covering engineering and also, for example, medicine and economics. Therefore, the terms digital filtering, Kalman filtering, and other types of processing, appear quite often in the present-day professional literature. The aim of this two-part book is to give a relatively simple introduction to digital and Kalman filtering. The first part covers the filtering operation as normally understood in electrical engineering and specified in the frequency-domain. The second part deals with the filtering of noisy data in order to extract the signal from noise in an optimum (minimum mean-square error) sense.

Digital filtering, as used in the title, refers to part 1 of the book, but the subtitle specifies it more closely as an introduction to discrete-time filtering which is a common theory for digital or other types of filter realizations. The actual realization of discrete-time filters, not discussed here, can be done either in sampled-data form (using bucket-brigade or charge coupled devices), or in digital form (using binary logic circuitry). Alternatively, discrete-time filters, described in terms of difference equation algorithms, can be handled on digital computers.

Kalman filtering, as used in the title, refers to part 2 of the book, and again the subtitle describes it more closely as an introduction to linear estimation theory developed in discrete-time domain. Although this part deals initially with some digital filter structures, it develops its own terminology. It introduces the criterion of the minimum mean-square error, scalar and vector Wiener and Kalman filtering. However, the main and most practical topic is the Kalman filtering algorithm which in most applications requires the use of digital computers.

Most of the material has been used by the author in postgraduate courses over the past five years. The presentation is in 'tutorial' form, but readers are assumed to be familiar with basic circuit theory, statistical averages, and elementary matrices. Various central topics are developed gradually with a number of examples and problems with solutions. Therefore, the book is suitable both for introductory postgraduate and undergraduate courses.

The author wishes to acknowledge helpful discussions with Dr J. A. Edwards of this department, who also helped with some of the computer programs.

SMB
1979

Part 1 – Digital filtering

Introduction

Digital filtering is used here as a well-established title, but with the reservation that we are dealing only with time sequences of sampled-data signals. However, the fundamental theory presented for discrete-time signals is general and can also be applied to digital filtering.

It is important to clarify the terminology used here and in the general field of signal processing. The analogue or continuous-time signal means a signal continuous in both time and amplitude. However, the term continuous-time implies only that the independent variable takes on a continuous range of values, but the amplitude is not necessarily restricted to a finite set of values, as discussed by Rabiner (1). Discrete-time implies that signals are defined only for discrete values of time, i.e. time is quantized. Such discrete-time signals are often referred to as sampled-data or analogue sample signals. The widely-used term digital implies that both time and amplitude are quantized. A digital system is therefore one in which a signal is represented as a sequence of numbers which take on only a finite set of values.

It is also important to clarify the notation used here. In mathematics the time increments or decrements are denoted by Δt, but in digital filtering the sampling time interval T is generally used. A sample of input signal at time $t = kT$ is denoted as $x(k)$, where T is neglected (or taken as unity) and k is an integer number; similarly, for the output we have $y(k)$. In many papers and textbooks, particularly mathematical ones (difference equations), the notation is x_k, y_k. We use here the notation $x(k)$, $y(k)$ for the following reasons:

(i) it is a direct extension of the familiar $x(t)$, $y(t)$ notation used for the continuous-time functions;
(ii) it is suitable for extension to the state-variable notation, where the subscripts in $x_1(k)$, $x_2(k)$ . . . refer to states;
(iii) it is more convenient for handling complicated indices, for example $x(N-\frac{1}{2})$.

It will be seen later that digital filtering consists of taking (usually) equidistant discrete-time samples of a continuous-time function, or values of some discrete-time process, and performing operations such as discrete-time delay, multiplication by a constant and addition to obtain the desired result.

The first chapter introduces discrete-time concepts using some simple and familiar analogue filters, and also shows how a discrete-time description can arise directly from the type of operation of a system, e.g. radar tracking. The z-transform is then introduced as a compact representation of discrete-time sequences, and also as a link with the Laplace transformation. The second chapter expands the basic concepts established in the first chapter, dealing first with the time response of a discrete-time system (difference equations). Then we introduce and discuss the transfer function, inversion from z-variable back to time-variable, and the frequency response of a digital filter. This chapter ends with the realization schemes and classification of digital filters into the basic nonrecursive and recursive types, whose design techniques are presented in the third and fourth chapters respectively. In both cases design examples and computer calculated responses are given to illustrate various design methods. In the fifth chapter we return to some of the basic relationships introduced in the first chapter, and deal with two important topics in discrete-time processing. First we develop the discrete Fourier series representation of periodic sequences, which enables formulation of the discrete Fourier transform (DFT) for aperiodic finite sequences. The second topic is the inverse filter, which is an important concept used in many fields for removal or reduction of undesirable parts of a sequence. The concept of minimum error energy is introduced as an optimization technique for the finite length inverse filter coefficients. At the end of each chapter a number of problems is given with solutions at the end of the book.

It is of interest to mention that the approach used in this book is a form of transition from the continuous- (or analogue) to discrete-time (or digital) systems, since most students have been taught electrical engineering in terms of continuous-time concepts. An alternative approach is to study electrical engineering directly in terms of discrete-time concepts without a reference to continuous-time systems. It appears that the discrete case is a natural one to the uninhibited mind, but special and somewhat mysterious after a thorough grounding in continuous concepts, as discussed by Steiglitz (2), p. viii.

1

Introduction to discrete-time filtering

1.0 Introduction

In continuous-time, the filtering operation is associated with RC or LC type of circuits. Therefore, in the first section of this chapter, we consider two simple filtering circuits (RC and RLC) described in continuous-time by differential equations, and we find their discrete-time equivalents, i.e. difference equations. There are also situations in which difference equations are obtained directly, as illustrated in section 1.2. In continuous-time we usually represent differential equations in the complex frequency s-domain by means of Laplace transformation, and from these we obtain the frequency response along the $s = j\omega$ axis. Similarly, difference equations in discrete-time are transformed into z-domain using z-transformation which is briefly introduced in the third section of this chapter. The relationship between z and s is then established in section 1.4.

1.1 Continuous and discrete-time analysis

We are all familiar with the description of continuous-time dynamic systems in terms of differential equations. As an introduction to the discrete-time description and the process of filtering we consider a few differential equations and transform them into their discrete-time equivalents, i.e. difference equations.

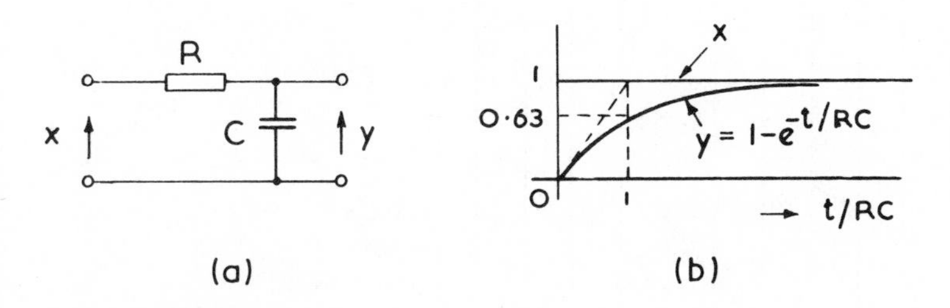

Fig. 1.1 (a) Simple first-order RC filter; (b) solution for unit step input

Consider the typical first-order RC filter in fig. 1.1(a) where x and y represent the input and output voltages respectively. For this simple network x and y are related by the differential equation

$$RC\frac{dy}{dt} + y = x \tag{1.1}$$

The solution, for a unit step input and zero initial condition, is shown in fig. 1.1(b). To derive the discrete-time equivalent for equation 1.1, we use the method of backward differences, described by Hovanessian *et al.* (3), and obtain

$$RC\frac{y_k - y_{k-1}}{\Delta t} + y_k = x_k$$

where we have used the standard mathematical notation. Solving for y_k, we have

$$y_k = \frac{1}{1+\Delta t/RC} y_{k-1} + \frac{\Delta t/RC}{1+\Delta t/RC} x_k$$

Using the approximation $(1+\Delta t/RC)^{-1} \simeq 1-\Delta t/RC$, where we have neglected terms of higher order, we obtain

$$y(k) = a_0\, x(k) + b_1\, y(k-1) \tag{1.2}$$

with $a_0 = \Delta t/RC$ and $b_1 = 1 - a_0$. Note that we have changed to sample notation, as discussed in the introduction, with k representing the discrete integer time parameter instead of $t = k\Delta t$. The above result enables us to draw fig. 1.2(a) which is the discrete-time equivalent

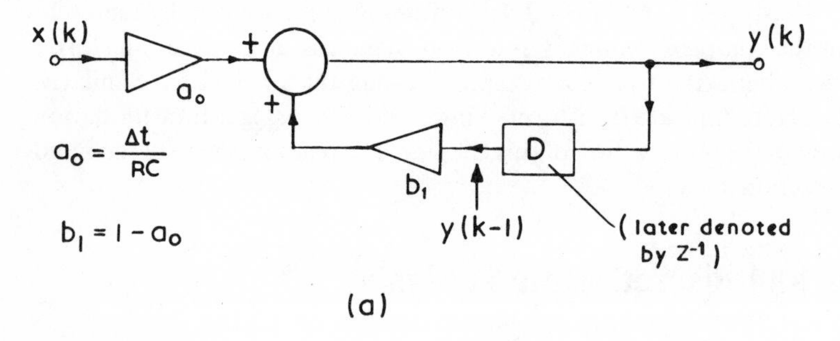

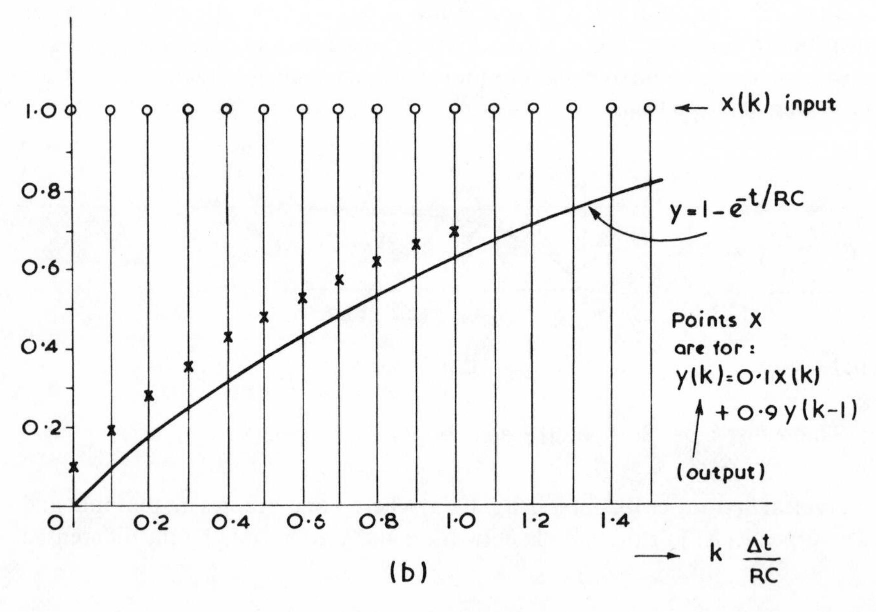

Fig. 1.2 (a) Discrete-time representation of fig. 1.1(a); (b) solution for unit step input

of the continuous-time system shown in fig. 1.1(a). In fig. 1.2(a), triangles are used to represent multiplication by the factor written beside them, rectangles to represent delay units denoted also by $D(=\Delta t)$, and circles to represent addition. Choosing numerical values $a_0 = \Delta t/RC = 0.1$, $b_1 = 1 - a_0 = 0.9$, and $y(-1) = 0$ as the initial condition, we obtain fig. 1.2(b), where the first ten points have been calculated using a slide rule.

A typical second-order differential equation is

$$\frac{d^2y}{dt^2} + 2\sigma\frac{dy}{dt} + \omega_0^2 y = \omega_0^2 x \tag{1.3}$$

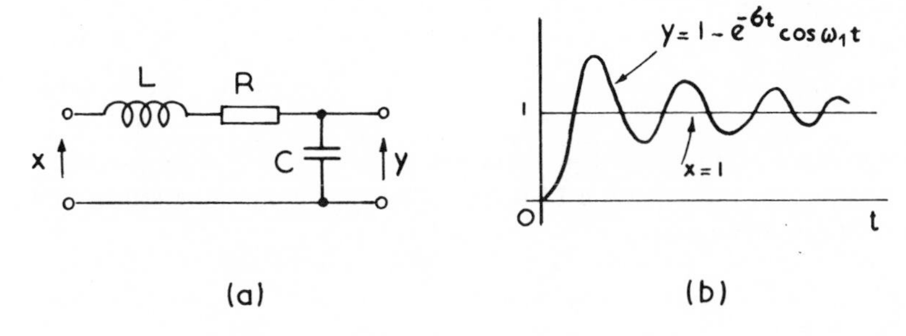

Fig. 1.3 (a) LRC filter; (b) solution for unit step input

where again x and y refer to the input and output respectively. Such an equation describes, for example, the LRC circuit in fig. 1.3(a) with $\sigma = R/2L$ and $\omega_0^2 = 1/LC$. The solution for the unit step input is given in fig. 1.3(b) for the underdamped case when $\omega_1 = \sqrt{(\omega_0^2 - \sigma^2)} \gg 1$. The discrete-time form of equation 1.3 can be shown to be

$$y(k) = a_0\, x(k) + b_1\, y(k-1) + b_2\, y(k-2) \tag{1.4}$$

where backward differences have been used, and the coefficients are functions of σ and ω_0 (see section 1 of the appendix). The discrete-time operational scheme for the second-order difference equation 1.4 is shown in fig. 1.4, with the same notation as in fig. 1.2(a). The unit

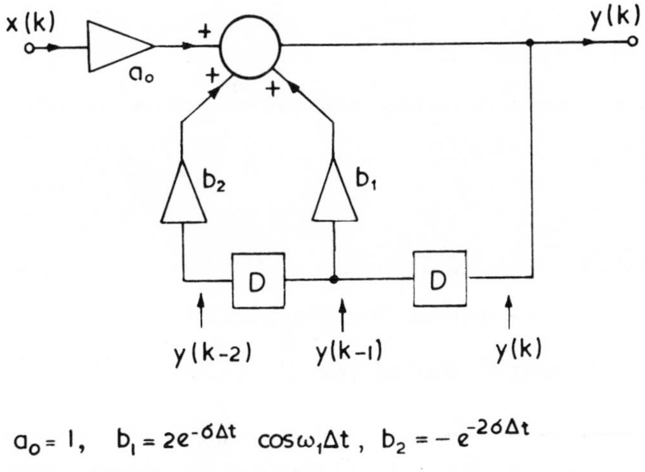

$a_0 = 1, \quad b_1 = 2e^{-\sigma\Delta t} \cos\omega_1\Delta t, \quad b_2 = -e^{-2\sigma\Delta t}$

Fig. 1.4 Discrete-time representation of fig. 1.3(a)

step response in discrete-time has not been calculated, but the interested reader may do it as an exercise.

The input $x(t)$ and output $y(t)$ variables in equations 1.1 and 1.3 are related in terms of differential equations. Another way of expressing their relationship, in continuous-time systems, is in terms of the convolution integral

$$y(t) = \int_0^t h(\tau)\,x(t-\tau)\,\mathrm{d}\tau \tag{1.5}$$

where $h(\tau)$ represents the impulse response, and $x(t)$ is an arbitrary input signal. The discrete-time form of the above equation is

$$y(k) = \sum_{i=0}^{k} h(i)\,x(k-i) \tag{1.6}$$

known as the convolution summation, often written as $y = h * x$ (note that $h * x = x * h$). We can illustrate in a simple way that equation 1.6 is correct by considering equation 1.2 first for the impulse (i.e. unit-sample) input,

$$x(k) = \delta(k) = \begin{cases} 1 & k = 0 \\ 0 & k \neq 0 \end{cases}$$

and then for an arbitrary input $x(k)$. In both cases we assume zero initial conditions.

For the impulse input, equation 1.2 produces the sequence

$$\begin{aligned} y(0) &= h(0) = a_0 \\ y(1) &= h(1) = a_0 b_1 \\ y(2) &= h(2) = a_0 b_1^2 \\ &\vdots \\ y(k) &= h(k) = a_0 b_1^k \end{aligned} \tag{1.7}$$

but for an arbitrary input signal $x(k)$, we obtain

$$\begin{aligned} y(0) &= a_0 x(0) \\ y(1) &= a_0 x(1) + a_0 b_1 x(0) \\ y(2) &= a_0 x(2) + a_0 b_1 x(1) + a_0 b_1^2 x(0) \\ &\vdots \\ y(k) &= a_0 x(k) + a_0 b_1 x(k-1) + \ldots + a_0 b_1^k x(0) \end{aligned}$$

Comparing the above equation with the set of values in equation 1.7 we have

$$y(k) = h(0)\,x(k) + h(1)x(k-1) + \ldots + h(k)x(0)$$

or $\quad y(k) = \sum_{i=0}^{k} h(i)x(k-i)$

which is the convolution summation given by equation 1.6.

1.2 Discrete-time processing in radar tracking

This is an example of direct formulation of the problem in discrete-time, and a simplified diagram, fig. 1.5(a), shows the main points to be discussed. A radar beam is used to determine the range and velocity of an object at a distance x from the transmitter. Fig. 1.5 shows the set

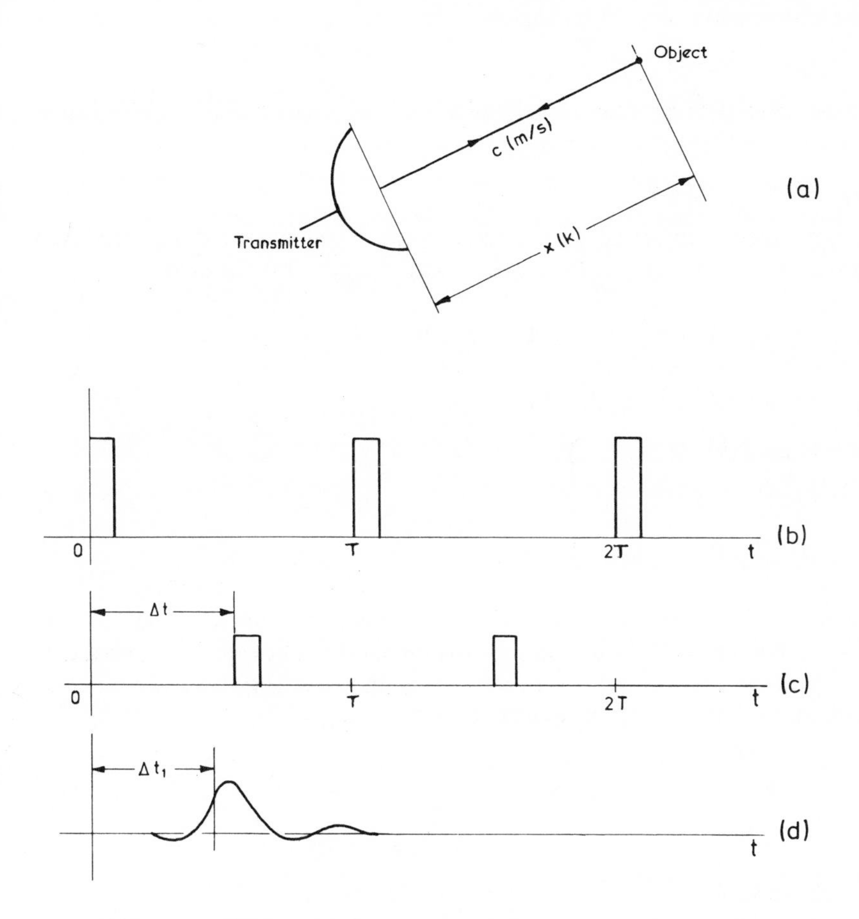

Fig. 1.5 (a) Simplified radar tracking system; (b) ideal transmitted pulses; (c) ideal received pulses; (d) typical received pulses

of ideal transmitted and received pulses together with a typical received pulse. The information required is the value of the time interval Δt representing the time passed for the radio wave to travel to the object and back. The typical received signal is not of ideal shape due to various disturbances, and we measure $\Delta t_1 \neq \Delta t$. The range estimate $x = c\,\Delta t_1/2$ from one measurement can therefore cause large errors; c is the speed of pulse propagation in space. To reduce the error, a periodic sequence of pulses is transmitted every T seconds, as indicated in fig. 1.5, which produces a sequence of measured values of range $x(0), x(1), \ldots, x(k)$. In many cases the object is moving and its velocity (rate of change of range) is required, together with the object's range at a time one radar pulse in the future.

To establish a processing scheme for radar data we introduce the following quantities:

$x(k)$, the measurement of the object's range obtained from the kth radar pulse return;
$y(k)$, the estimate of the object's range at the kth radar pulse after data processing;
$\dot{y}(k)$, the estimate of the object's velocity at the kth radar pulse after data processing;
$y_p(k)$, the prediction of the object's range at the kth radar pulse, obtained at the $(k-1)$th radar pulse after data processing.

The last of the above quantities can be expressed as

$$y_p(k) = y(k-1) + T\dot{y}(k-1)$$

where T is the time interval between transmitted pulses. The next relationship is established in the following way:

$$y(k) = y_p(k) + \alpha[x(k) - y_p(k)]$$

where the predicted range is corrected by the error between the measured and predicted values, which is scaled by factor $\alpha > 0$. In a similar way, we have for the velocity

$$\dot{y}(k) = \dot{y}(k-1) + \frac{\beta}{T}[x(k) - y_p(k)] \qquad \text{where } \beta > 0$$

The set of relationships formed from the above equations

$$\left.\begin{aligned} y_p(k) &= y(k-1) + T\dot{y}(k-1) \\ y(k) &= y_p(k) + \alpha[x(k) - y_p(k)] \\ \dot{y}(k) &= \dot{y}(k-1) + \frac{\beta}{T}[x(k) - y_p(k)] \end{aligned}\right\} \tag{1.8}$$

describes a signal processing scheme known as the alpha–beta (α–β) tracking equations. The hardware structure defined by these equations is shown in fig. 1.6. The input $x(k)$ represents the measured data, and the three outputs of the processing unit, $y(k)$, $y(k)$ and $y_p(k)$ represent the range, velocity and range prediction respectively.

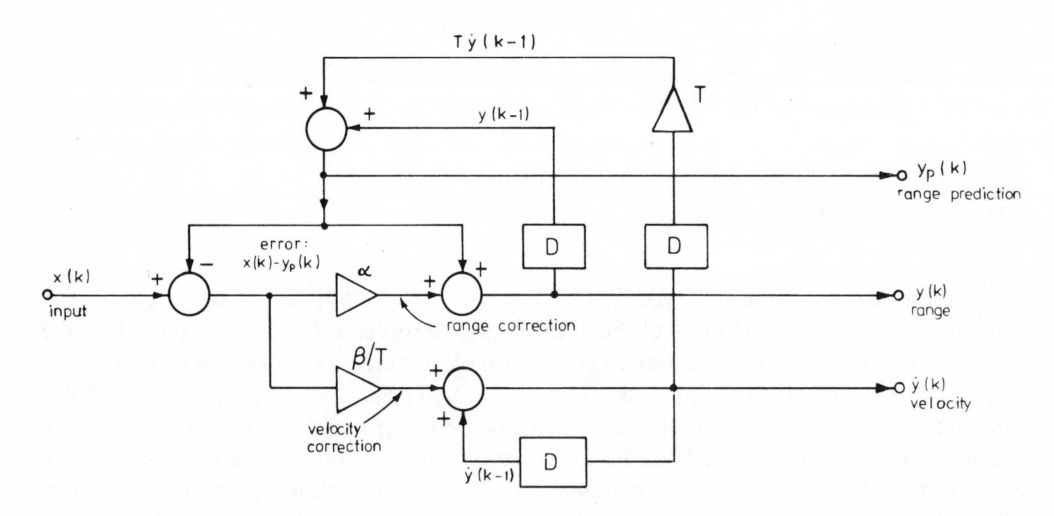

Fig. 1.6 α–β tracker processor

More detail on the α–β tracking can be found in Cadzow (4), but the above set of equations is further considered in problems 1.6 and 1.7, and later in problem 2.14.

The block diagrams shown in figs 1.2(a), 1.4 and 1.6 are digital filters or processors represented in the hardware form. Their computational algorithms, given by equations 1.2, 1.4 and 1.8 respectively, are often used as such in time series analysis, for example in

Anderson(5). However, in engineering practice it is usually preferable to work in the frequency-domain, not only because such representation shows clearly the type of filtering, but because design and measurements are often more precise in the frequency-domain than in the time-domain.

We know that in continuous-time systems the time-domain is transformed into the complex frequency (s-) domain using the Laplace transform. Similarly, in discrete-time systems we can use the z-transform which enables analysis of these systems in the frequency-domain. An introduction to z-transform and its properties is given in the following two sections.

1.3 *z*-transform

Let $f(k)$ represent any discrete-time variable, i.e. $x(k)$ or $y(k)$ in the examples of sections 1.1 and 1.2, or any other discrete-time variable obtained for example from data recording. This discrete-time variable is a time sequence which can be written, for example, as

$$[f(k)] = f(0), f(1), f(2), \ldots, f(k), \ldots$$

taking place at times

$$0, \Delta t, 2\Delta t, \ldots, k\Delta t, \ldots$$

A useful way of representing this data is in the form of a polynomial

$$f(0) + f(1)z^{-1} + f(2)z^{-2} + \ldots + f(k)z^{-k}$$

which is a power series in z^{-k} having as coefficients the values of the time sequence $[f(k)]$. Alternatively, this can be written in the following compact form (see, for example, Cadzow (4) and Freeman (6)):

$$Z[f(k)] = F(z) = \sum_{k=0}^{\infty} f(k)z^{-k} \tag{1.9}$$

where we have assumed $f(k) = 0$ for $k < 0$. Such a function is known as the z-transform of the $f(k)$ sequence. The z-transform gives a procedure by which a sequence of numbers can be transformed into a function of the complex variable z. The variable z^{-k} can be interpreted as a type of operator that upon multiplication shifts signal samples to the right (delays) by k time units. Multiplication by z^k shifts signals to the left. Another interpretation of the variable z is developed in the following section, where it is shown that z, in sampled-data systems, is a complex number such that $z = e^{sT}$, where T ($= \Delta t$ in the notation used up to here) is the sampling time interval.

There are various methods of finding closed form expressions for the z-transforms of common time sequences, which are discussed by Cadzow(4) and Freeman(6). As an illustration, we consider the case of the geometric sequence (or exponential signal) given by $f(k) = c^k$ for $k \geq 0$, and $f(k) = 0$ for $k < 0$. The z-transform of this sequence is given by

$$Z[f(k)] = F(z) = \sum_{k=0}^{\infty} c^k z^{-k}$$

This is a geometric series whose sum is given by

$$S = A\frac{1 - r^k}{1 - r}$$

where $r = cz^{-1}$, $A = 1$, and $k \to \infty$, and we can write

$$Z[c^k] = \frac{1}{1 - cz^{-1}} \tag{1.10}$$

for $|r| = |cz^{-1}| < 1$ or $|z| > |c|$. The following three cases are typical sequences whose z-transforms can be obtained from equation 1.10. First, for $c = 1$, we have

$$Z[1] = \frac{1}{1 - z^{-1}} \tag{1.11}$$

for $|z| > 1$, which is the z-transform of the constant sequence (or sampled unit step). Similarly, by letting $c = ae^{jb}$, we have

$$Z[a^k e^{jbk}] = \frac{1}{1 - ae^{jb} z^{-1}} \tag{1.12}$$

for $|z| > a$. From equation 1.12 we can easily obtain $Z[a^k \cos bk\,\Delta t]$ and $Z[a^k \sin bk\,\Delta t]$, as in Table 1.1.

Furthermore, we can obtain the z-transform for the sampled ramp signal $x(k) = k$, for $k \geq 0$, from equation 1.10 as follows. Writing equation 1.10 as

$$\sum_{k=0}^{\infty} c^k z^{-k} = \frac{z}{z - c}$$

and differentiating both sides with respect to z,

$$z^{-1} \sum_{k=0}^{\infty} k\, c^k z^{-k} = \frac{c}{(z - c)^2}$$

Multiplying by z and setting $c = 1$, we have

$$\sum_{k=0}^{\infty} k\, z^{-k} = \frac{z}{(z - 1)^2} \tag{1.13}$$

which is also given in Table 1.1.

We consider next the case of a signal delayed by i discrete-time units. Its z-transform is given by

$$Z[f(k - i)] = \sum_{k=0}^{\infty} f(k - i)\, z^{-k}$$

or

$$Z[f(k - i)] = z^{-i} \sum_{k} f(k - i)\, z^{-(k - i)}$$

$$= z^{-i} \sum_{m=-i}^{\infty} f(m)\, z^{-m}$$

which results in

$$Z[f(k - i)] = z^{-i} F(z) \tag{1.14}$$

where we have assumed zero initial conditions $f(m) = 0$ for $m < 0$ (for nonzero conditions, see, for example, Cadzow(4), pp. 162–4). This is an important result, also known as the right-shifting property, which will be used later in various sections. The above relationship corresponds to the Laplace time-shift theorem

$$L[f(t-\tau)] = e^{-\tau s} F(s) \tag{1.15}$$

Derivations of many other z-transform relationships can be found in Cadzow(4), Freeman(6) and Jury(7). In Table 1.1 we have used T in the last two entries instead of Δt. This is the notation which is generally accepted and which will be used from the next section onwards.

Table 1.1 Commonly used z-transforms

$x(k),\ k \geq 0$		$X(z)$
Unit pulse	$\delta(k)$	1
Unit step	1	$1/(1-z^{-1})$
Exponential	c^k	$1/(1-cz^{-1})$
Ramp	k	$z^{-1}/(1-z^{-1})^2$
Cosine wave	$a^k \cos bkT$	$(1-az^{-1}\cos bT)/(1-2az^{-1}\cos bT + a^2z^{-2})$
Sine wave	$a^k \sin bkT$	$az^{-1}\sin bT/(1-2az^{-1}\cos bT + a^2z^{-2})$

We shall be using the z-transform as defined by equation 1.9. However, mathematicians and some control engineers use the Laplace definition of z-transform (described by Robinson(8), p. 341) given by

$$F(z) = \sum_{k=0}^{\infty} f(k) z^k$$

1.4 z-transform relation to the Laplace transform

In communications and control practice, the discrete-time sequences $x(k)$ and $y(k)$, or in the general notation $f(k)$, are samples of a continuous-time waveform. These can be interpreted as a form of modulation of the signal $f(t)$ by a sequence of impulses as in fig. 1.7. A signal $f(t)$ is sampled every T seconds, and the sampled output wave

$$f_s(t) = f(t)\, x f_1(t)$$

can be represented as

$$f_s(t) = f(t) \sum_{k=0}^{\infty} \delta(t-kT)$$

or $$f_s(t) = \sum_{k=0}^{\infty} f(kT)\,\delta(t-kT) \tag{1.16}$$

where the subscript s denotes the sampled signal. (In control systems analysis, sampled waveforms are commonly denoted by an asterisk.) The notation in this section is changed from the one used previously in two ways: $f(k)$ is changed to $f(kT)$, and Δt is changed to T. This notation is widely used in the field of digital filters.

The Laplace transform of $\delta(t-kT)$ is

$$L[\delta(t-kT)] = e^{-skT}$$

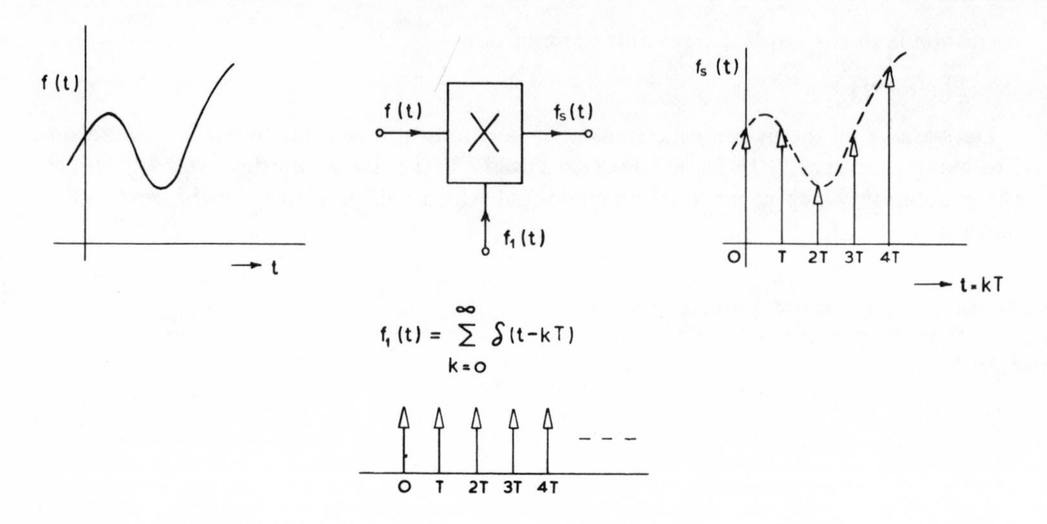

Fig. 1.7 Modulation of a signal by a sequence of impulses

Therefore, the Laplace transform of the sampled waveform described by equation 1.16 is given by

$$F_s(s) = \sum_{k=0}^{\infty} f(kT)\mathrm{e}^{-skT} \tag{1.17}$$

Comparing equation 1.17 with equation 1.9 and taking into account the change in notation discussed above, we have

$$F(z) = F_s(s)\Big|_{\mathrm{e}^{sT}=z} \tag{1.18}$$

which establishes the relationship between z and complex frequency s, i.e.

$$z = \mathrm{e}^{sT} \tag{1.19}$$

This is an important relationship and it is examined in some detail in section 2 of the appendix.

There may be differences in the treatment of equation 1.16 and those following from it. For example, Kaiser (9) multiplies equation 1.16 by T and defines the z-transform as $TF(z)$, where $F(z)$ is given by equation 1.9. The factor T is explained as an approximation of $f_s(t)$ to $f(t)$ in the sense that the area under both functions is approximately the same in the interval $kT < t < (k+1)T$, provided that T is sufficiently small. It seems to be more correct to include T in equation 1.16 and other related expressions, since a multiplier such as the cutoff frequency ω_c then becomes $\omega_c T$ which is dimensionally correct (see section 4.1). We shall keep equation 1.17 as shown since we have defined $F(z)$ by equation 1.9, which is the commonly-used form.

The Laplace transform, equation 1.17, enabled us to link s-plane with z-plane as expressed by equation 1.19. The importance of this relationship will be seen in the following chapters. There is another form of Laplace transform of the sampled wave (section 3 of the appendix) given by

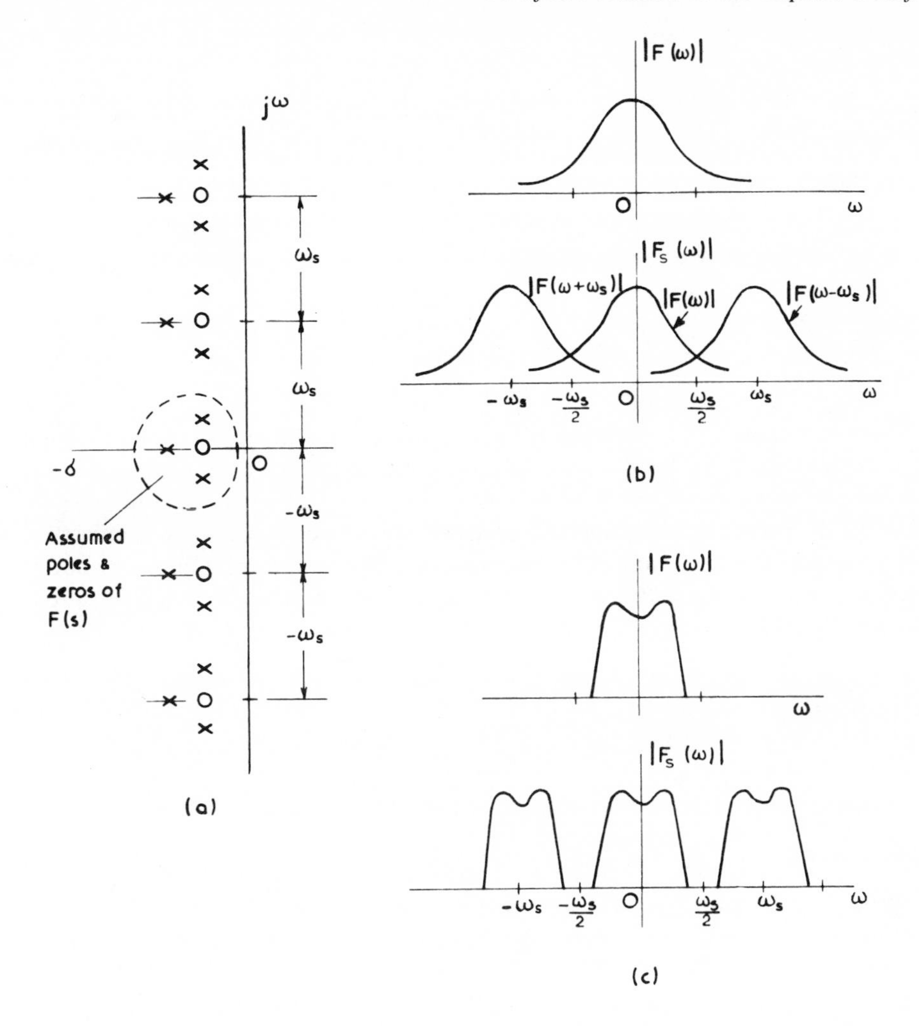

Fig. 1.8 (a) Pole zero pattern for a sampled signal, where ω_s is the sampling frequency; (b) frequency spectra before sampling $|F(\omega)|$, and after sampling $|F_s(\omega)|$; (c) $|F(\omega)|$ and $|F_s(\omega)|$, but $|F(\omega)|$ is well bandlimited to frequencies below $\omega_s/2$

$$F_s(s) = \frac{1}{T}\sum_{n=-\infty}^{\infty} F(s + jn\omega_s) \tag{1.20}$$

where $\omega_s = 2\pi/T$ represents the sampling frequency (rad s^{-1}), and $F(s)$ is the Laplace transform of the input signal. The roots of $F_s(s)$ are periodic in the s-plane as shown in fig. 1.8(a), and so is $F_s(s)$ as shown in figs 1.8(b) and (c). The main point of fig. 1.8(b), also expressed in fig. 1.8(c), is that in order to avoid interspectra interference (or frequency aliasing), the input signal must be bandlimited to frequencies below $\omega_s/2$. As will be seen later, the digital filter design is confined to the frequency range $-\omega_s/2 \leq \omega \leq \omega_s/2$. The required filter characteristic (lowpass, bandpass, etc.) is then specified within this frequency range.

The general scheme of continuous-signal simulation in discrete-time form is shown in

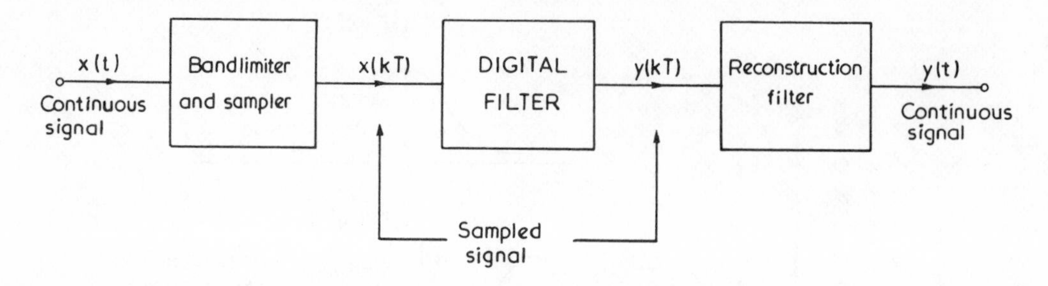

Fig. 1.9 Continuous-signal processing in discrete-time form

fig. 1.9. The bandlimiter before the sampler is a good lowpass filter to restrict the input signal to frequencies below $\omega_s/2$. Such a filter should have a flat magnitude response and a linear phase over the major portion of the interval $\pm\omega_s/2$, and a large attenuation at frequencies greater than $\omega_s/2$. A minimal sampling rate is usually best since it results in a minimum number of computational operations. The reconstruction filter is a lowpass filter with cutoff frequency of $\omega_s/2$, producing a continuous-time signal from the time sequence of output samples of the digital filter.

1.5 Problems

1.1 Show that the difference equation for the highpass connection of the circuit in fig. 1.1(a) is given by

$$y(k) = (1 - \Delta t/RC)[y(k-1) + x(k) - x(k-1)]$$

1.2 (a) Write the z-transform for the finite sequence given by $[1, 0, 1/4, 0, 1/16]$.
(b) If the above sequence extends to infinity, i.e. $[1, 0, 1/4, 0, 1/16, 0, 1/64, 0, \ldots]$, write its z-transform.

1.3 (a) Plot the exponential signal $x(k) = c^k$ over the range $k = 0, 1, 2, \ldots, 10$, for $c = \pm 0.5, \pm 1.5$.
(b) The z-transform for the above signal is given in Table 1.1. Plot its pole-zero positions in the z-plane, for the values of c given in part (a). Compare the results of (a) and (b), and show why some cases represent bounded (stable) signals while the others unbounded (unstable) signals; see, for example, section 2 of the appendix.

1.4 Find the z-transform of $x(k) = 1 + k$, and plot the pole-zero pattern.

1.5 Write z-transforms for equations 1.2 and 1.4 given in section 1.1.

1.6 Apply z-transform to the set of equations 1.8 and show that

$$Y(z) = \frac{z(\alpha z + \beta - \alpha)}{D} X(z)$$

$$\dot{Y}(z) = \frac{\beta}{T} \frac{z(z-1)}{D} X(z)$$

$$Y_p(z) = \frac{z(\alpha + \beta) - \alpha}{D} X(z)$$

where $D = z^2 + (\beta + \alpha - 2)z + (1 - \alpha)$.

1.7 Assuming that in equations 1.8 $x(k) = \delta(k)$ is a unit pulse, show that, for the equations in problem 1.6 to represent a critically damped (nonoscillatory) system, the relationship between α and β must satisfy

$$\alpha = 2\sqrt{\beta} - \beta$$

1.8 Show that the final value $f(\infty)$ is obtained from

$$f(\infty) = \lim_{z \to 1} (z - 1)F(z) \tag{1.21}$$

Note: This only applies for cases in which $(z - 1)F(z)$ is analytic for $|z| \geq 1$ (see, for example, Cadzow (4)).
Hint: Start with

$$Z[f(k+1) - f(k)] = \lim_{N \to \infty} \sum_{k=0}^{N} [f(k+1) - f(k)]z^{-k}$$

1.9 (a) Apply z-transform to equation 1.6 and show that

$$Y(z) = H(z)X(z)$$

where $H(z) = \sum_{i=0}^{k} h(i)z^{-i}$

(b) Let x and h be finite length sequences each consisting of two terms: $x = (2, 1)$ and $h = (3, 4)$. Determine their convolution sequence firstly from $y = h * x$ (see equation 1.6), and secondly from $Y(z) = H(z)X(z)$.

2

Digital filter characterization

2.0 Introduction

The basic analytical tools and relationships which have been introduced so far enable us to develop the concepts similar to analogue filters. In the first section of this chapter, we derive the digital filter transfer function $H(z)$ in the z-domain, and explore its properties by calculating the impulse response and frequency response. Several simple examples are used to illustrate methods of analysis.

We also discuss the digital filter hardware structures and consider their relative merits. The last section deals with the classification of digital filters into nonrecursive and recursive types, and the general properties of these two classes are briefly discussed.

2.1 Digital transfer function

We have already seen that in a linear discrete-time system, the input $x(k)$ and output $y(k)$ sequences are related by linear difference equations with constant coefficients as in equations 1.2 and 1.4. In the theory of digital or discrete-time filters, the general difference equation is usually written in the following way:

$$y(k)+b_1y(k-1)+\ldots+b_My(k-M) = a_0x(k)+a_1x(k-1)+\ldots+a_Nx(k-N) \qquad (2.1)$$

where b_0 is taken, by convention, as unity.

The interpretation of equation 2.1 is that at time $k(t = kT)$, the output value can be computed from the current input and a linear combination of previous inputs and outputs. The output sequence, or response of a digital filter for a given input sequence, can therefore be calculated in a simple manner as illustrated below in example 2.1.

Example 2.1
Taking $a_0 = 1$, $a_n = 0$ for $n \neq 0$, and $b_1 = \pm 0.8$, $b_m = 0$ for $m \geq 2$, equation 2.1 reduces to

$$y(k) = x(k) \mp 0.8y(k-1) \qquad (2.2)$$

If the input is a unit pulse described as $x(0) = 1$ and zero for all $k \neq 0$, and assuming initial conditions to be $x(k) = y(k) = 0$ for $k < 0$, we can form the calculation scheme shown in Table 2.1. The signs $\mp$ refer to filters with $\pm b_1 = \pm 0.8$. The input and output sequences for this case are shown in fig. 2.1, where the full lines are for $b_1 + 0.8$ and also dashed ones for $b_1 = -0.8$.

If the input is a unit step given by $x(k) = 1$ for $k \geq 0$, with the same initial conditions as in the above case, we obtain Table 2.2. The input and output sequences for this case are shown in fig. 2.2. The final values y(∞) are also given in figs 2.1 and 2.2. They have been calculated using

Table 2.1

k	$x(k)$	$y(k)$	$y(k-1)$
0	1	1	0
1	0	∓0·80	1
2	0	0·64	∓0·80
3	0	∓0·51	0·64
4	0	0·41	∓0·51
5	0	∓0·33	0·41
6	0	0·26	∓0·33

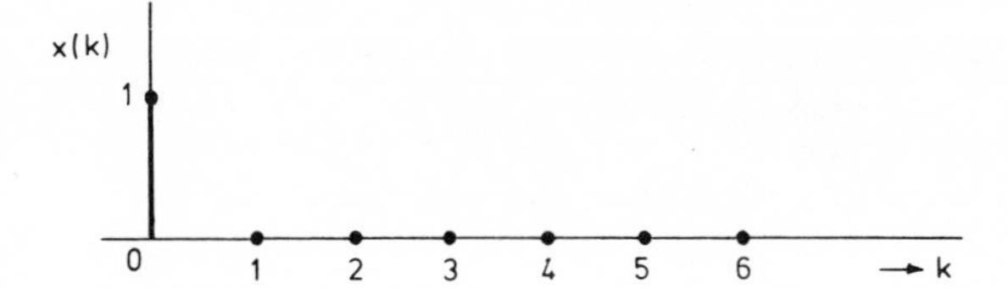

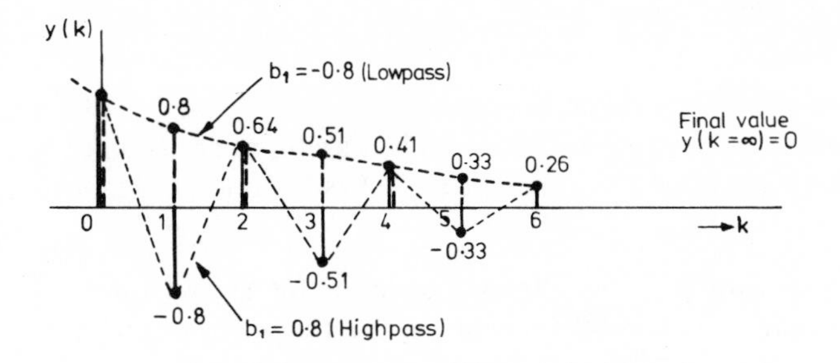

Fig. 2.1 Filter response for the unit pulse input

Table 2.2

		$b_1 = 0{\cdot}8$		$b_1 = -0{\cdot}8$	
k	$x(k)$	$y(k)$	$y(k-1)$	$y(k)$	$y(k-1)$
0	1	1	0	1	0
1	1	0·20	1	1·8	1
2	1	0·84	0·20	2·44	1·8
3	1	0·33	0·84	2·95	2·44
4	1	0·74	0·33	3·35	2·95
5	1	0·41	0·74	3·69	3·35
6	1	0·67	0·41	3·95	3·69

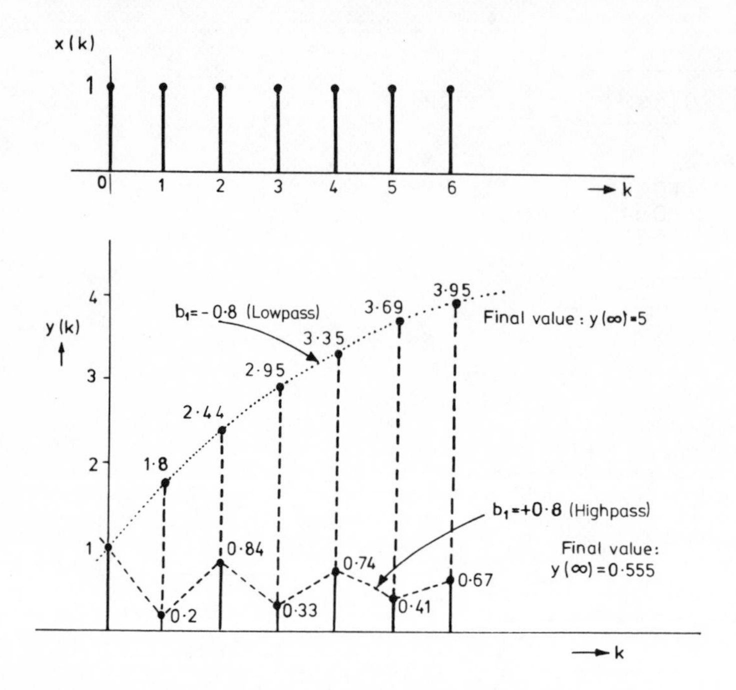

Fig. 2.2 Filter response for the unit step input

the final value equation 1.21 established in problem 1.8; $y(\infty)$ is a useful quantity since it shows whether the output sequence converges. We shall refer to figs 2.1 and 2.2 again in example 2.4.

The above illustrated step-by-step solution of the difference equation for a given input is a straightforward process, but if we require a closed-form solution and frequency response of the output sequence, we need an alternative approach based on the *z*-transform.

Taking the *z*-transform of equation 2.1 term-by-term, and using equations 1.9 and 1.14 as appropriate, we obtain

$$Y(z)\left(1+\sum_{m=1}^{M} b_m z^{-m}\right) = X(z)\sum_{n=0}^{N} a_n z^{-n} \tag{2.3}$$

where $X(z) = \sum_{k=0}^{\infty} x(k)z^{-k}$ and $Y(z) = \sum_{k=0}^{\infty} y(k)z^{-k}$

From equation 2.3 we can now define the discrete-time (or digital) transfer function as

$$H(z) = \frac{Y(z)}{X(z)} = \frac{\sum_{n=0}^{N} a_n z^{-n}}{1+\sum_{m=1}^{M} b_m z^{-m}} \tag{2.4}$$

which is a rational polynomial in z^{-1}. This transfer function is valid for zero initial conditions which are satisfied in the cases we are considering. For nonzero initial conditions, see, for example, Cadzow (4), p. 227.

We have seen that the *z*-transformation of the difference equation 2.1 enabled us to define the digital transfer function. This is analogous to the Laplace transform of a differential equation in continuous-time, which produces the transfer function $H(s)$ in s-domain. From equation 2.4 we can write for the output

$$Y(z) = H(z)X(z) \tag{2.5}$$

and the output sequence $y(k)$ is then obtained using the inverse z-transform. The special case of equation 2.5 is obtained for the unit pulse input sequence

$$x(k) = \begin{cases} 1 & k = 0 \\ 0 & k \neq 0 \end{cases}$$

which has z-transform $X(z) = 1$. The response to this input is then the inverse z-transform of $H(z)$. Therefore, the input sequence $[1, 0, 0, \ldots]$ in discrete-time filter theory corresponds to the unit impulse in continuous-time filter theory, and the inverse z-transform of $H(z)$ is the impulse response of a digital filter. We note here that the impulse function in discrete-time, as defined above, is a more realistic quantity than the corresponding impulse function in continuous-time.

2.2 The inverse transformation

To obtain the impulse response sequence $h(k)$ from $H(z)$, or $y(k)$ from $Y(z)$, we can use one of the three methods of inversion:

(i) inversion by series expansion (long division);
(ii) inversion by partial-fraction expansion;
(iii) use of an inversion integral described by Freeman (6) and Oppenheim & Schaffer (10).

We illustrate the first two methods of inversion using the filter of example 2.1.

Example 2.2
Applying z-transform to equation 2.2 we obtain the following transfer function:

$$H(z) = \frac{1}{1 \pm 0.8z^{-1}} = \frac{z}{z \pm 0.8} \tag{2.6}$$

The inversion of $H(z)$, by direct division of the numerator of the above equation by the denominator using long division, produces the series

$$H(z) = 1 \mp 0.8z^{-1} + 0.64z^{-2} \mp 0.51z^{-3} + 0.41z^{-4} \mp 0.33z^{-5} + 0.26z^{-6} + \ldots$$

The coefficients in this expansion are the values of the impulse response: $h(0) = 1$, $h(1) = \mp 0.8$, $h(2) = 0.64$, $h(3) = \mp 0.51$, etc. We note that these coefficients are the same as $y(k)$ values in Table 2.1.

The series expansion is useful when the first few terms of the sequence $h(k)$ are to be found. The general solution, in closed form, is obtained using partial-fraction expansion. The inversion of $H(z)$ by partial-fractions, in this case, is obtained by recognising that $H(z)$ in equation 2.6 is the third entry in Table 1.1. We then have immediately $h(k) = \mp 0.8^k$, which is the closed form solution of the above obtained impulse response series.

Example 2.3
The unit step input has the z-transform

$$X(z) = \frac{1}{1 - z^{-1}} \tag{2.7}$$

given by the second entry in Table 1.1. The z-transformed output is then obtained from equation 2.5 for the filter, with $b_1 = 0.8$ specified by equation 2.6, as

$$Y(z) = \frac{z^2}{z^2 - 0.2z - 0.8} \tag{2.8}$$

Dividing the numerator by the denominator we obtain the following series:

$$Y(z) = 1 + 0.2z^{-1} + 0.84z^{-2} + 0.33z^{-3} + 0.738z^{-4} + 0.412z^{-5} + 0.672z^{-6} + \dots$$

whose coefficients represent the filter output sequence for the unit step input. Note that these values are the same as in Table 2.2.

For the partial-fraction method we have to determine the poles which, for equation 2.8, are found to be 1 and -0.8. Then we rewrite equation 2.6 as

$$Y(z) = z\left[\frac{z}{(z-1)(z+0.8)}\right] \tag{2.9}$$

The function inside the square bracket has the degree of its numerator less than the degree of its denominator. Such rational functions are called proper rational functions, and they can be represented as a sum of simple one-pole terms as follows:

$$\frac{z}{(z-1)(z+0.8)} = \frac{A}{z-1} + \frac{B}{z+0.8} \tag{2.10}$$

The constants A and B can be found in the following manner. Multiply both sides of equation 2.10 by $(z-1)$, and setting $z = 1$ we have

$$A = \left.\frac{z}{z+0.8}\right|_{z=1} = \frac{1}{1.8}$$

Now multiplying both sides of equation 2.10 by $z + 0.8$, and setting $z = -0.8$, we obtain

$$B = \left.\frac{z}{z-1}\right|_{z=-0.8} = \frac{1}{2.25}$$

Therefore, equation 2.9 can be written as

$$Y(z) = \frac{1}{1.8}\left(\frac{z}{z-1}\right) + \frac{1}{2.25}\left(\frac{z}{z+0.8}\right)$$

or

$$Y(z) = \frac{1}{1.8}\left(\frac{1}{1-z^{-1}}\right) + \frac{1}{2.25}\left(\frac{1}{1+0.8z^{-1}}\right) \tag{2.11}$$

The factors in brackets are the second and third entries in Table 1.1, from which we obtain corresponding time sequences resulting in the output sequence

$$y(k) = \frac{1}{1.8} + \frac{1}{2.25}(-0.8)^k \tag{2.12}$$

This is the closed form solution which can be checked for $k = 0, 1, 2, 3, \dots$ with the coefficients obtained in the earlier series expansion.

In the above example poles of $Y(z)$ have been simple-order poles. A similar method applies to multiple-pole cases, and is described by Cadzow (4) and Stanley (11).

2.3 Frequency response

With regard to filtering we are particularly interested in the interpretation of the transfer (or system) function $H(z)$ as a frequency selective function. Therefore, we consider difference equation 2.1 when the input is a sampled complex exponential waveform

$$x(k) = e^{jk\omega T} \quad \text{for } k = 0, 1, 2, 3, \ldots \tag{2.13}$$

We can take then that the output is

$$y(k) = F(\omega)\, e^{jk\omega T} \tag{2.14}$$

since the system is linear, and hence the output is of the same frequency as the input, with $F(\omega)$ being a complex proportionality factor depending on ω only and determined as follows.

Using equations 2.13 and 2.14 in equation 2.1, we have

$$e^{jk\omega T} F(e^{j\omega T})(1 + b_1 e^{-j\omega T} + b_2 e^{-j2\omega T} + \ldots + b_M e^{-jM\omega T})$$
$$= e^{jk\omega T}(a_0 + a_1 e^{-j\omega T} + a_2 e^{-j2\omega T} + \ldots + a_N e^{-jN\omega T})$$

and
$$F(e^{j\omega T}) = \frac{\sum_{n=0}^{N} a_n e^{-jn\omega T}}{1 + \sum_{m=0}^{M} b_m e^{-jm\omega T}} \tag{2.15}$$

Comparing equation 2.15 with equation 2.4, we see that

$$F(e^{j\omega T}) = H(z)\Big|_{z = e^{j\omega T}} = H(e^{j\omega T}) \tag{2.16}$$

Thus the frequency response of a discrete-time filter is determined by the values of its transfer function on the unit circle in the z-plane given by $|z| = |e^{j\omega T}| = 1$. This is again analogous to the frequency response of a continuous-time filter being determined by the values of its transfer function on the imaginary axis. The s–z plane correspondence is examined in more detail in section 2 of the appendix.

Example 2.4
To determine the frequency response of the filter described by the transfer function 2.6, we set $z = e^{j\omega T}$ and obtain

$$H(\omega) = \frac{1}{1 \pm 0.8e^{-j\omega T}}$$

Expanding $e^{-j\omega T}$ in the usual manner, and calculating the magnitude response we have

$$|H(\omega)| = \frac{1}{1.28\sqrt{(1 \pm 0.975 \cos \omega T)}}$$

Evaluating this function at points $\omega T = 0, \pi/4, \pi/2, 3\pi/4, \pi$, we obtain the result plotted in fig. 2.3. The abscissa is marked in angle ωT and frequency ω. It is seen that $|H(\omega)|$ is a periodic function of ω. However, it is sufficient to specify the filter characteristics in the frequency range $0 \leq \omega \leq \omega_s/2$, i.e. up to half of the sampling frequency ($\omega_s/2 = \pi/T$) as discussed in section 1.4, and particularly fig. 1.8(c).

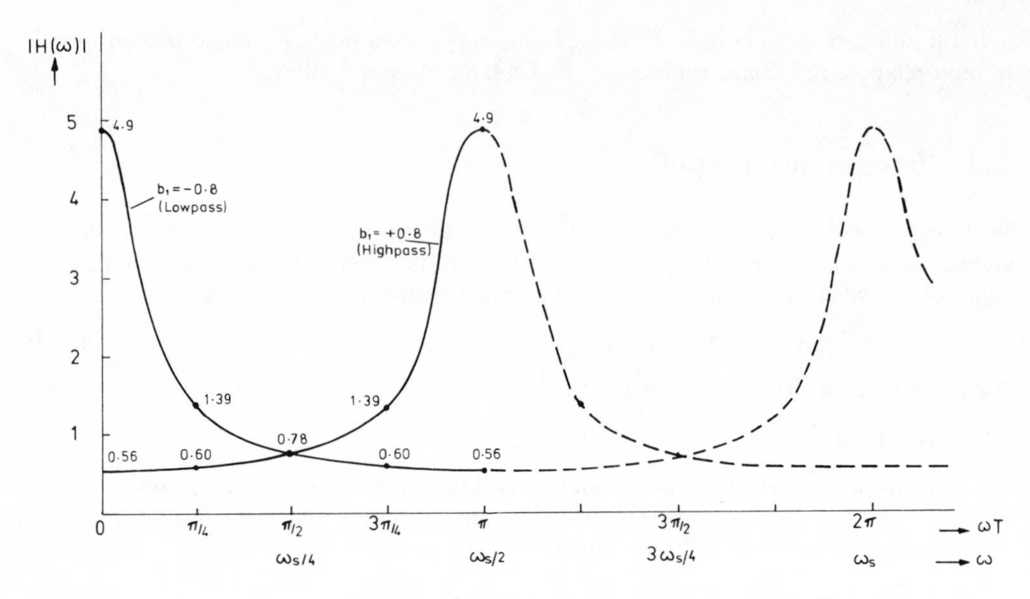

Fig. 2.3 Frequency response of the filter specified by equation 2.6

The graph for $b_1 = +0.8$ shows that the filter is a highpass in the frequency range $0 \le \omega \le \omega_s/2$, but for $b_1 = -0.8$, we have a lowpass filter in the same frequency range. Therefore, as expected, the frequency selective properties of the filter are clearly shown in the frequency-domain. These properties are not as clearly shown in the time-domain graphs in figs 2.1 and 2.2, but even there, one can observe that the output of the filter with $b_1 = -0.8$ allows slow time variations (lowpass filter in figs 2.1 and 2.2), while the output of the filter with $b_1 = 0.8$ allows fast time variations to pass through (highpass filter in figs 2.1 and 2.2).

It is perhaps a suitable place here to point out that discrete-time processing of signals is much simpler to deal with than continuous-time processing, since finding the response of a difference equation to inputs such as impulse or step is a straightforward algebraic calculation which can be done in a short time. However, the corresponding time-domain analysis of analogue filters means that we have to solve a differential equation every time we change its order or the input signal.

2.4 Digital filter realization schemes

Digital filter realization of a given transfer function $H(z)$ is quite simple. Suppose the input is $x(k)$ with z-transform $X(z)$, and the output is $y(k)$ with z-transform $Y(z)$. Then, from equation 2.4, we have

$$Y(z) = \sum_{n=0}^{N} a_n z^{-n} X(z) - \sum_{m=1}^{M} b_m z^{-m} Y(z)$$

and applying equations 1.9 and 1.14 in inverse sense, we obtain the time-domain equation

$$y(k) = \sum_{n=0}^{N} a_n x(k-n) - \sum_{m=1}^{M} b_m y(k-m) \tag{2.17}$$

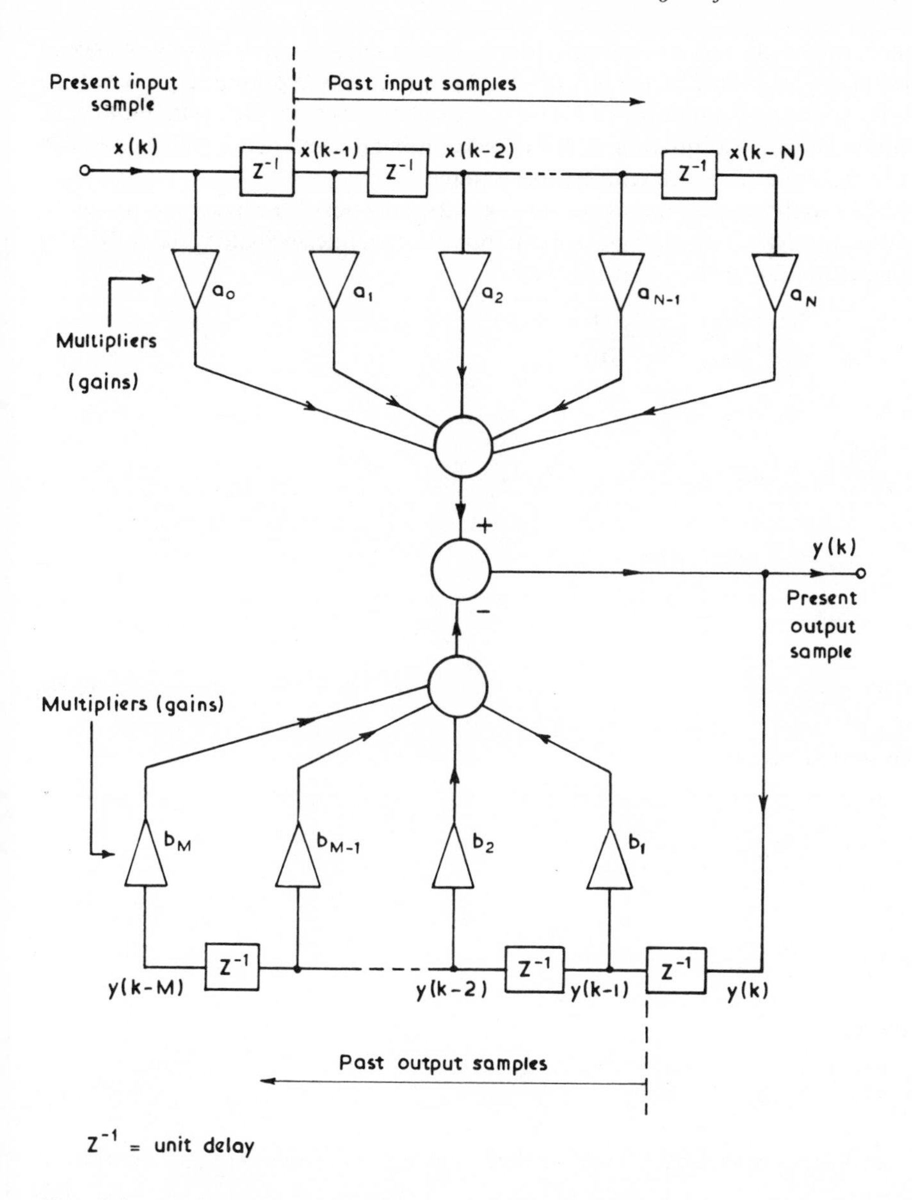

Fig. 2.4 Hardware implementation of a discrete-time filter

which is the difference equation that realizes $H(z)$ directly, and it also represents a computational algorithm. A hardware implementation of equation 2.17 is shown in fig. 2.4. As before, in figs 1.2(a) and 1.4, a triangle represents multiplication of a variable by the constant written beside it, a rectangle represents a unit-sample delay, and a circle represents a summing point. The interpretation of fig. 2.4 is as follows: at $t = kT$, $x(k)$ becomes available and the quantities $x(k-1), x(k-2), \ldots, x(k-N), y(k-1), \ldots, y(k-M)$ at the outputs of the delay elements have been remembered, i.e. stored. Thus all the variables are available for the computation of $y(k)$. When this computation is complete $x(k-N)$ and $y(k-M)$ are discarded, but the other quantities are saved since they will be needed for the next computation, i.e. input $x(k+1)$ and output $y(k+1)$. In this way, an entire input sequence of infinite duration can be filtered by the algorithm 2.17 to produce an output sequence of the same length. By counting the number of delays in fig. 2.4, we obtain a minimum number of storages required in realizing equation 2.17, and by counting the number of triangles we can see how many multiplications are required per sample.

Equation 2.17 and the corresponding fig. 2.4 is not the only possible way to realize a given digital filter function $H(z)$. For example, we can introduce an intermediate variable $W(z)$ by partitioning equation 2.4 in the following way:

$$H(z) = \frac{Y(z)}{X(z)} = \frac{Y(z)}{W(z)} \cdot \frac{W(z)}{X(z)} = N(z)\frac{1}{D(z)}$$

The first part is

$$N(z) = \frac{Y(z)}{W(z)} = \sum_{n=0}^{N} a_n z^{-n}$$

leading to $\quad Y(z) = \sum_{n=0}^{N} a_n z^{-n} W(z)$

or in time-domain

$$y(k) = \sum_{n=0}^{N} a_n w(k-n) \tag{2.18}$$

Similarly the second part is

$$\frac{1}{D(z)} = \frac{W(z)}{X(z)} = \frac{1}{1 + \sum_{m=1}^{M} b_m z^{-m}}$$

leading to

$$W(z) = X(z) - \sum_{m=1}^{M} b_m z^{-m} W(z)$$

or in time-domain

$$w(k) = x(k) - \sum_{m=1}^{M} b_m w(k-m) \tag{2.19}$$

Therefore, in place of equation 2.17 we have the following set of computational algorithms:

$$\left.\begin{aligned} w(k) &= x(k) - \sum_{m=1}^{M} b_m w(k-m) \\ y(k) &= \sum_{n=0}^{N} a_n w(k-n) \end{aligned}\right\} \tag{2.20}$$

with the hardware implementation shown in fig. 2.5. An advantage of equations 2.20 over equations 2.17 is a reduction in memory requirement, since we need to save only N or M previous values of $w(k)$, depending on which is greater. This is illustrated in fig. 2.6 (redrawn fig. 2.5 with $N = M$) in which the delay elements having the same output have been replaced by a single delay element. This is known as the canonic form because it has the minimum number of delay elements, but since other configurations also have this property, this terminology is not recommended (see Rabiner (1), p. 328).

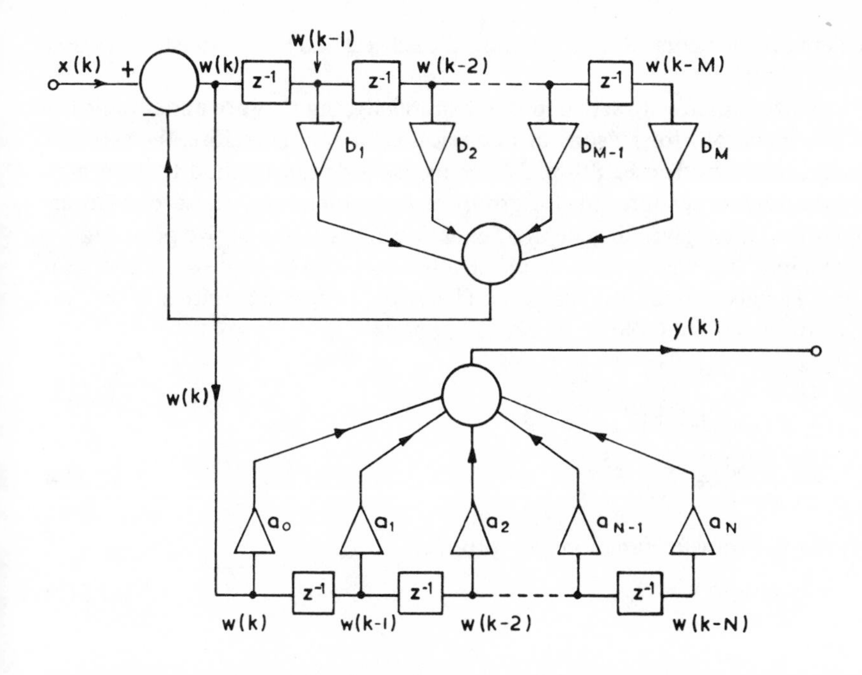

Fig. 2.5 Alternative hardware implementation of discrete-time filter

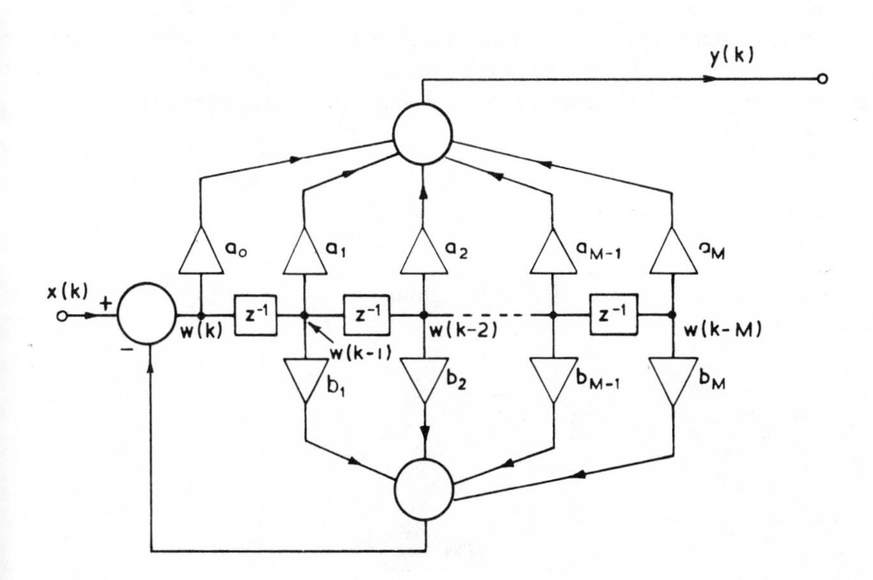

Fig. 2.6 Representation of fig. 2.5 with a reduced number of delay elements (canonic form)

The realizations of digital filters shown in figs 2.4 and 2.6 are called direct forms. It is interesting to note that the constants a_n and b_m in the network, i.e. hardware implementation, are the same as the constants in the transfer function. In continuous-time systems this is not the case since the network elements (R, L, C) are not so easily related to the transfer function of a continuous-time filter.

The direct form realizations are attractive for their simplicity, but they are undesirable for high-order difference equations for reasons of numerical accuracy. Therefore, the cardinal rule to follow, as described by Kaiser (9), p. 272, is to use a design method that permits decomposition of the high-order filter into a group of low-order (first- or second-order) subfilters. The high-order filter, given in general by equation 2.4, is a ratio of two polynomials in z^{-1} or, if required, in z. It is well known that such a function can be expressed either as a product or as a sum of a partial-fraction expansion. Therefore, we may have either of the two following forms, as discussed by Gold & Rader (12), pp. 44–5.

$$H(z) = \frac{\prod_{n=1}^{N_1}(z^2 + c_n z + d_n)\ \prod_{n=1}^{N_2}(z + g_n)}{\prod_{m=1}^{M_1}(z^2 + e_m z + f_m)\ \prod_{m=1}^{M_2}(z + l_m)} \tag{2.21a}$$

$$\text{or}\quad H(z) = \sum_{n=1}^{N_1} \frac{z(A_n z + B_n)}{z^2 + C_n z + D} + \sum_{n=1}^{N_2} \frac{zE_n}{z + G_n} \tag{2.22a}$$

which lead to writing the transfer function as a product

$$H(z) = H_1(z) \times H_2(z) \times \ldots \times H_k(z) \tag{2.21b}$$

or as a sum

$$H(z) = H_1(z) + H_2(z) + \ldots + H_k(z) \tag{2.22b}$$

Equations 2.21 and 2.22 have been written in terms of poles and zeros of $H(z)$ which must lie on the real axis or occur in complex-conjugate pairs. The product form of equation 2.21b can be physically interpreted as a serial or cascade connection of first- and second-order subfilters as in fig. 2.7. The form of equation 2.22b can be interpreted as a parallel arrangement of first- and second-order subfilters as in fig. 2.8. Each of the subfilters can be realized in either of the direct forms discussed earlier.

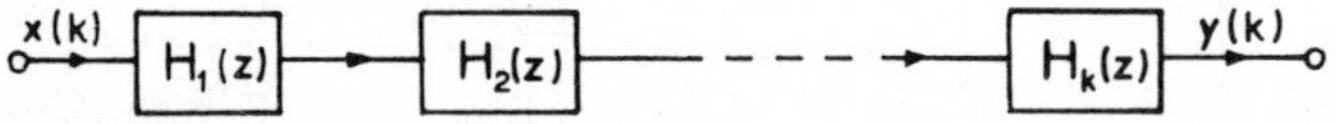

Fig. 2.7 Serial connection of first- and second-order subfilters

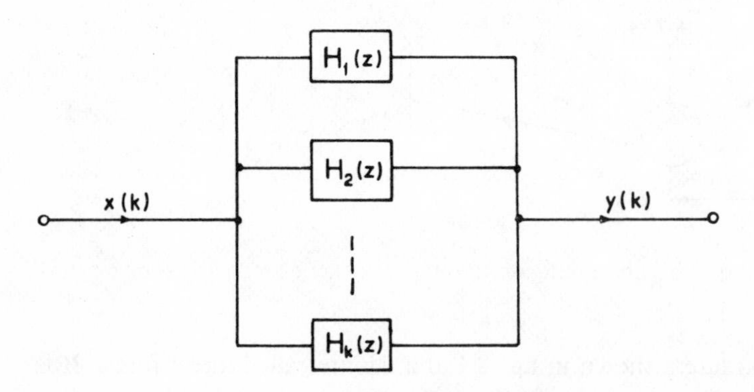

Fig. 2.8 Parallel connection of first- and second-order subfilters

2.5 The classification of digital filters

For the purpose of realization, digital filters are classified into nonrecursive and recursive types. The nonrecursive structure contains only the feed-forward paths as shown in fig. 2.9. This is a special case of equation 2.4 in which all b_m coefficients are zero, i.e. the output is a sum of linearly weighted present and a number of previous samples of the input signal. In recursive filter structures the output depends both on the input and on the previous outputs, as shown in the general hardware realization of fig. 2.4, where we have both feed-forward and feed-back paths.

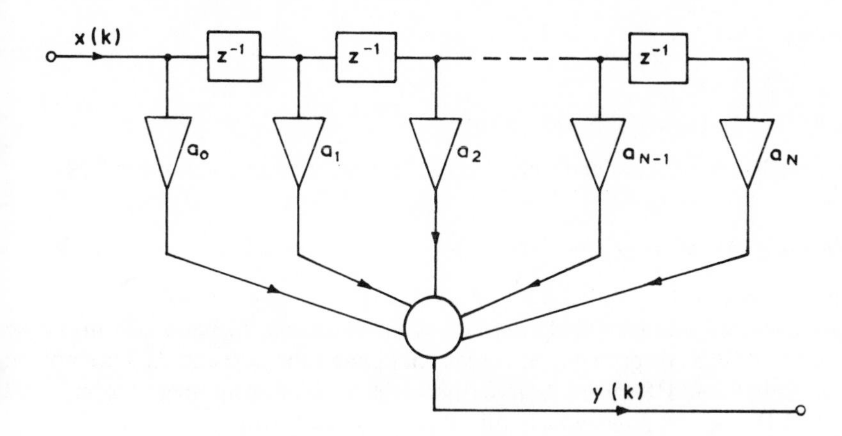

Fig. 2.9 Nonrecursive discrete-time filter

An alternative division of digital filters is made, on the basis of the impulse response duration, into finite impulse response (FIR) and infinite impulse response (IIR) filters. The simplest FIR filter realization is in the nonrecursive form. For example, a nonrecursive filter

$$y(k) = x(k) + a_1 x(k-1) \tag{2.23}$$

has a finite impulse response given by $[1, a_1, 0, 0, \ldots]$, but the simplest IIR filter realization is in the recursive form, for example,

$$y(k) = x(k) - b_1 y(k-1) \tag{2.24}$$

with impulse response $[1, -b_1, b_1^2, -b_1^3, b_1^4, \ldots]$, which has an infinite number of terms.

The easy implementation of FIR and IIR filters, in terms of the nonrecursive and recursive forms respectively, gives an impression of identity between them. However, in general both FIR and IIR filters can be implemented by either nonrecursive or recursive techniques. To illustrate this, we consider first the IIR filter of equation 2.24 for which we can write the transfer function as

$$H(z) = \frac{1}{1 + b_1 z^{-1}} \tag{2.25}$$

Series expansion of $H(z)$ produces

$$H(z) = 1 - b_1 z^{-1} + b_1^2 z^{-2} - b_1^3 z^{-3} + \ldots \tag{2.26}$$

and the corresponding difference equation can be written as

$$y(k) = x(k) - b_1x(k-1) + b_1^2x(k-2) - b_1^3x(k-3) + \ldots$$

$$\text{or} \quad y(k) = \sum_{i=0}^{k} (-b_1)^i x(k-i) \tag{2.27}$$

which is clearly a nonrecursive realization of the IIR filter given by equation 2.24.

Applying a similar procedure to the FIR filter of equation 2.23, we have the transfer function

$$H(z) = 1 + a_1z^{-1} \tag{2.28}$$

which can be written as

$$H(z) = \frac{1}{(1 + a_1z^{-1})^{-1}} = \frac{1}{1 - a_1z^{-1} + a_1^2z^{-2} - a_1^3z^{-3} + \ldots} \tag{2.29}$$

where the denominator has been expanded into a series. The difference equation for 2.29 is given by

$$y(k) = x(k) + a_1y(k-1) - a_1^2y(k-2) + a_1^3y(k-3) - \ldots \tag{2.30}$$

which represents a recursive realization of the FIR filter given by equation 2.23.

Although we have obtained mathematical alternatives for equations 2.23 and 2.24 in the form of equations 2.30 and 2.27 respectively, it is seen that equations 2.30 and 2.27 are not desirable realizations of FIR and IIR filters. A large (theoretically infinite) number of terms is required to obtain the FIR and IIR filters which follow quite naturally from the nonrecursive and recursive realizations respectively. Equations 2.27 and 2.30 can also be derived directly from equations 2.24 and 2.23 respectively, without using z-transform. This is the object of problem 2.12.

It is of interest to note also that in time series analysis the nonrecursive filter is known as the moving-average filter, and the recursive filter is the autoregressive moving-average filter, as discussed by Anderson (5).

2.6 Problems

2.1 For a filter described by the linear difference equation

$$y(k) = x(k) = x(k) + 0.5y(k-1)$$

determine the first five terms of its response to the input

$$x(k) = \begin{cases} 0 & k < 0 \\ 1,\ -2,\ 2,\ 1,\ -2 & k = 0,\ 1,\ 2,\ 3,\ 4, \text{ respectively.} \end{cases}$$

Consider two different initial conditions: (a) $y(-1) = 0$ and (b) $y(-1) = 2$. Solve this problem in time-domain.

2.2 Solve the problem specified above in problem 2.1 using z-transform. Compare the result obtained with the solution of problem 2.1 for $y(-1) = 0$.

2.3 Solve the problem specified in problem 2.1, with the initial condition $y(-1) = 2$, using z-transform. Compare the result with the solution in problem 2.1.

Note: In this case use modified input z-transform given by

$$\overline{X}(z) = X(z) + 0.5y(-1)$$

2.4 Derive a general expression for the response of the filter

$$y(k) = a_0x(k) - b_1y(k-1)$$

to the input signal

$$x(k) = \begin{cases} 0 & k < 0 \\ 1 & k = 0, 2, 4, 6, \ldots \\ -1 & k = 1, 3, 5, 7, \ldots \end{cases}$$

The initial condition is $y(-1) = 0$. Develop the solution by working in time-domain.

2.5 Determine the first five terms of the inverse transform of the following two functions:

$$X_1(z) = \frac{z^{-1} + z^{-2}}{1 - z^{-2}}, \qquad X_2(z) = \frac{1 + 2z^{-1}}{1 - \frac{1}{2}z^{-1}}$$

(a) by long division and (b) by partial-fraction expansion.

2.6 Determine the first five terms of the inverse transform of the function

$$X(z) = \frac{1}{1 - z^{-1} + \frac{1}{2}z^{-2}}$$

(a) by long division and (b) by partial-fraction expansion.

2.7 (a) Determine the difference equation relating $x(k)$ and $y(k)$ for the system shown in fig. 2.10. Write its transfer function $H(z)$.
(b) Using the transfer function, determine the first six terms of the impulse response of this system.
(c) Derive the amplitude and phase response of this system. What kind of filtering does it perform?

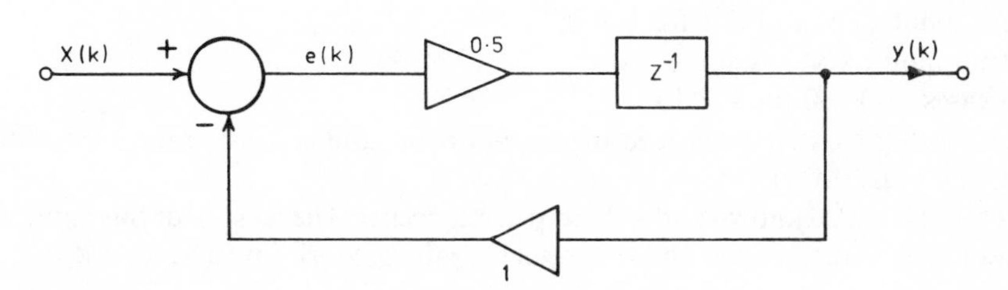

Fig. 2.10

2.8 Examine the effect on the filter

$$y(k) = \tfrac{1}{2}x(k) + \tfrac{1}{2}x(k-1)$$

if, in its transfer function, z is replaced by z^2. More specifically, how are the amplitude and phase response affected? Plot simple graphs of these quantities as functions of ωT.

2.9 In radar systems a moving target indicator employs a filter to remove the background clutter, as described fully by Skolnik (13). The simplest type of such a filter is the single delay line canceller shown in fig. 2.11 as Unit 1. Calculate the amplitude and phase response of this unit (output y_1), and also for the cascade of two such units (output y). Compare the results by sketching simple graphs at points $\omega T = 0, \pi/2, \pi$.

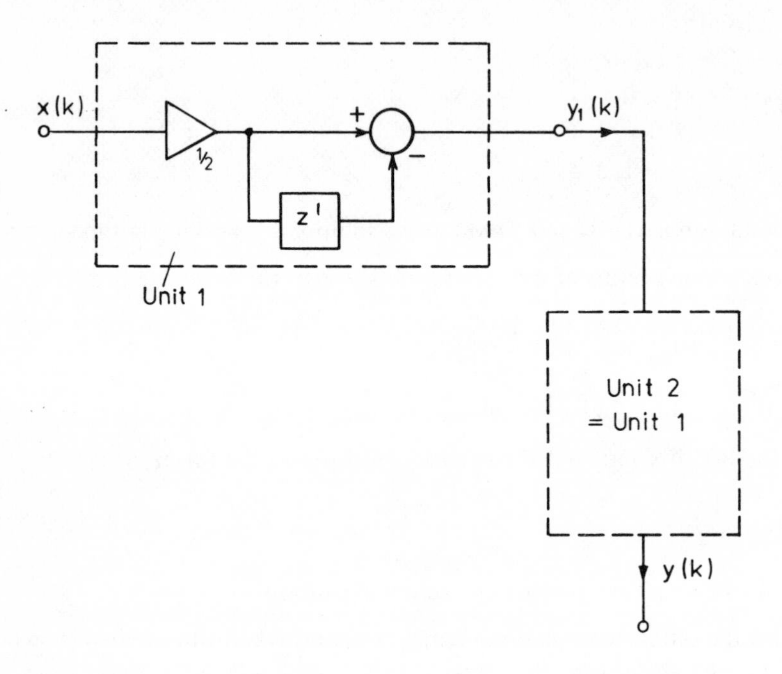

Fig. 2.11

2.10 Calculate the output sequence of the double delay canceller (cascade of the two units in fig. 2.11 of problem 2.9), for:

(a) unit input step, $x(k) = 1$ for $k \geq 0$;

(b) ramp input, $x(k) = k$ for $k \geq 0$.

In both cases $x(k) = 0$ for $k < 0$.

Perform a step-by-step analysis in the time-domain, and also using the z-transform method, $Y(z) = H(z)X(z)$.

Note: This case is also known as the three-pulse canceller. The reason for this name will be found in the course of the above specified analysis; work up to at least $k = 5$.

2.11 (a) Determine the difference equation relating $x(k)$ and $y(k)$ for the system shown in fig. 2.12.

(b) Derive the frequency response and calculate the amplitude response at $\omega T = 0, \pi/2$ and π. Sketch the graph passing through these points and compare it with the graphs for the amplitude response of the canceller in problem 2.9.

2.12 Derive equations 2.27 and 2.30 from equations 2.24 and 2.23 respectively. Work in the time-domain using a step-by-step approach; assume $y(-1) = 0$ and $x(-1) = 0$.

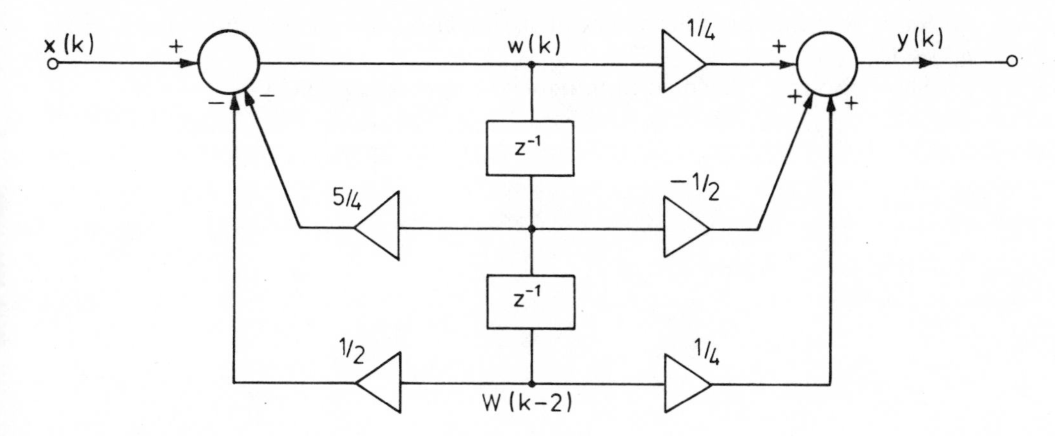

Fig. 2.12

2.13 A simple FIR filter is described by the difference equation

$$y(k) = x(k) + \tfrac{1}{2}x(k-1)$$

Its recursive realization can be expressed as

$$y(k) = x(k) + \sum_{n=1}^{N} (-1)^{n+1}(\tfrac{1}{2})^n y(k-n)$$

as in equations 2.29 and 2.30.

Draw the structure of the above recursive filter and determine its impulse response up to $k = 14$ for the case of $N = 5$. Compare the results with the original FIR response $[1, \tfrac{1}{2}, 0, 0, 0, \ldots]$. Assume initial conditions $y(k) = 0$ for $k < 0$.

2.14 The system of equations governing α–β tracker, as given in problems 1.6 and 1.7, can be written in the following way:

$$H_1(z) = \frac{z[(2\sqrt{\beta} - \beta)z + 2\beta - 2\sqrt{\beta}]}{(z + \sqrt{\beta} - 1)^2}$$

$$H_2(z) = \frac{\beta}{T}\frac{z(z-1)}{(z + \sqrt{\beta} - 1)^2}$$

$$H_3(z) = \frac{(2\sqrt{\beta})z + \beta - 2\sqrt{\beta}}{(z + \sqrt{\beta} - 1)^2}$$

where $H_1 = Y/X_0$ $H_2 = \dot{Y}/X$ and $H_3 = Y_p/X$ are the three transfer functions.
(a) Show that for the tracker to have stable dynamics, β must be within the range $0 < \beta < 4$.
(b) Calculate and compare graphically the impulse responses of $H_1(z)$ for $\beta = 0.81$ and for $\beta = 3.6$.

2.15 The transfer function of a discrete-time filter is given by

$$H(z) = \frac{5z^2 - 12z}{z^2 - 6z + 8}$$

(a) Show that the first four values of its impulse response are $h(0) = 5$, $h(1) = 18$, $h(2) = 68$, $h(3) = 264$.
(b) Show also that the closed form impulse response is given by

$$h(k) = 2^k + 4^{k+1}$$

3

Nonrecursive filter design

3.0 Introduction

In the previous chapter we have further developed the analytical tools which enabled us to perform the time and frequency analysis of given digital filters, and we have classified filters and discussed their properties. The aim of this chapter and the next is to use the techniques developed so far and to produce systematic procedures for the design of digital filters. The division into nonrecursive and recursive types will be used in preference to FIR and IIR filter division, since FIR and IIR filters are naturally easy to realize as nonrecursive and recursive structures respectively.

The design methods for each of these two classes of filters are different because of their distinctly different properties. The nonrecursive filter has a finite memory and can have excellent linear phase characteristics, but it requires a large number of terms, possibly in the order of several hundred, to obtain a relatively sharp cutoff frequency response. However, program execution time for nonrecursive filters can be considerably reduced using the Fast Fourier Transform techniques given by Helms(14). The recursive filter has an infinite memory and tends to have fewer terms, but its phase characteristics are not as linear as the nonrecursive ones. Recursive digital filters usually meet the stringent specifications arising in practice with at most ten to twenty coefficients, so the computation required to produce each output, given a new input, is of the order of ten to twenty multiplications and additions per sample point.

Filter design can be implemented in the time-domain or frequency-domain. We shall be dealing with the design of filters specified in the frequency-domain because it is more precise and more suitable for electrical engineering practice. The time-domain approach may be useful to give an initial guess as described by Rabiner & Gold(15), p. 265, and the filter characteristic is then specified in the frequency range $0 \leq \omega \leq \omega_s/2$, where ω_s is the sampling frequency. This range is often referred to as the baseband. Typical idealized lowpass, highpass, bandpass and bandstop amplitude responses for only the positive frequency ranges are shown in fig. 3.1. It is seen that they are specified within the frequency range 0 to $\omega_s/2$. The other parts of the frequency spectrum are periodic repetitions of the baseband which should have no effect on the baseband range, as discussed in section 1.4.

The main design methods for nonrecursive and recursive filters are given in this and the following chapter. Each of the two classes is treated separately and suitable design methods are developed and illustrated. Other design methods which lie outside the scope of this book are described by Oppenheim & Shaffer(10) and Rabiner & Gold(15).

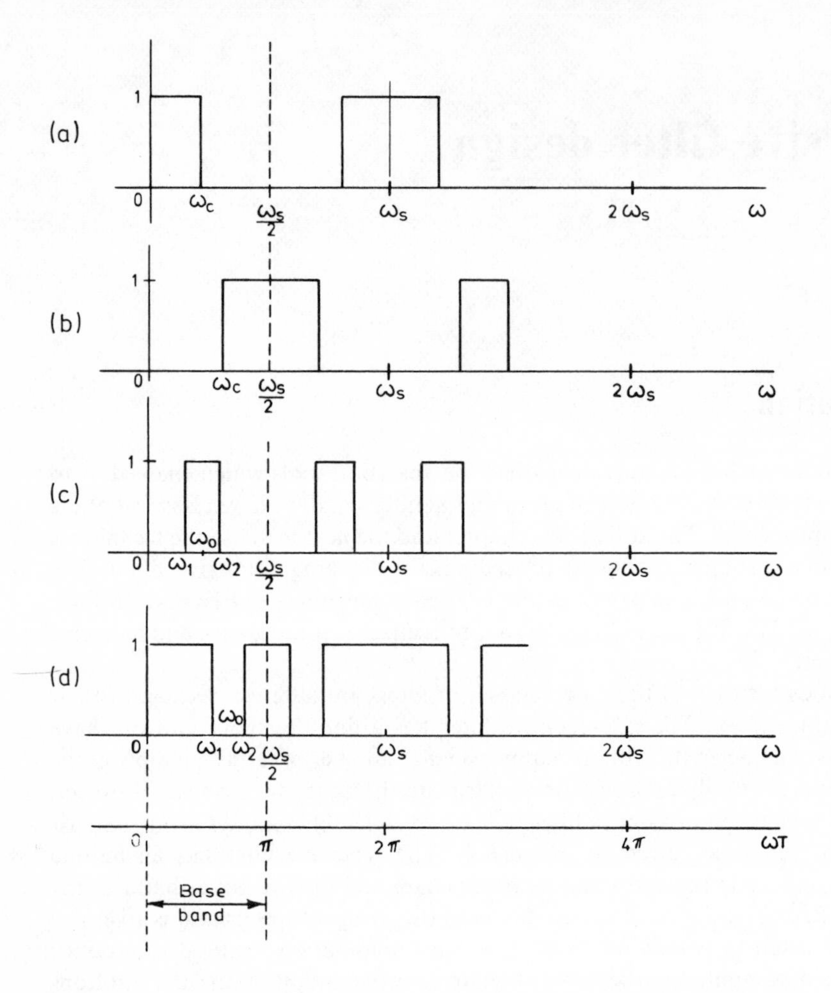

Fig. 3.1 Ideal amplitude responses of (a) lowpass, (b) highpass, (c) bandpass and (d) bandstop filters

3.1 Nonrecursive filter properties

The nonrecursive filter transfer function

$$H(z) = \sum_{n=0}^{N} a_n z^{-n} \tag{3.1}$$

follows from equation 2.4 for all $b_m = 0$. The corresponding difference equation is

$$y(k) = \sum_{n=0}^{N} a_n x(k-n) \tag{3.2}$$

from equation 2.17 for $b_m = 0$, or directly from equation 3.1 by expressing $H(z) = Y(z)/X(z)$ and applying z-transform rules of section 1.3.

We noted in chapter 2 that the coefficients a_n are impulse responses of the nonrecursive filter. To show this formally we set the input to be a unit pulse defined by

$$x(k) = \delta(k) = \begin{cases} 1 & k = 0 \\ 0 & k \neq 0 \end{cases}$$

Then the output is given by

$$y(k) = h(k) = \sum_{n=0}^{N} a_n \delta(k-n)$$

where $\delta(k-n) = \begin{cases} 1 & k = n \\ 0 & k \neq n \end{cases}$

Therefore, we have

$$h(k) = a_k \tag{3.3}$$

and $$H(z) = \sum_{n=0}^{N} h(n) z^{-n} \tag{3.4}$$

We also know that the transfer function $H(\omega) = H(z = e^{j\omega T})$ is a periodic function of frequency. Therefore, we can apply Fourier series to this function and obtain the coefficients $h(n)$. Before we express this relationship properly, it may be advantageous to remind ourselves of the familiar Fourier series for the periodic time functions. In this case the Fourier pair of equations, in complex form, is given by

$$f(t) = \sum_{n=-\infty}^{\infty} c_n e^{j2\pi nt/(T_2 - T_1)} \tag{3.5}$$

$$c_n = \frac{1}{T_2 - T_1} \int_{T_1}^{T_2} f(t) e^{-j2\pi nt/(T_2 - T_1)} \, dt \tag{3.6}$$

where $f(t)$ is the periodic time function with period $(T_2 - T_1)$ and c_n is a sequence of frequency-domain numbers, usually referred to as the frequency spectrum.

In digital filters the frequency response is a periodic function, so reversing the roles of time and frequency, we have

$$H(\omega) = \sum_n a_n e^{-j2\pi n\omega/(\Omega_2 - \Omega_1)} \tag{3.7}$$

$$a_n = \frac{1}{\Omega_2 - \Omega_1} \int_{\Omega_1}^{\Omega_2} H(\omega) e^{j2\pi n\omega/(\Omega_2 - \Omega_1)} \, d\omega \tag{3.8}$$

where $H(\omega)$ is a periodic frequency response function, and $\Omega_2 - \Omega_1$ is its period. The coefficients a_n form a time-domain sequence of numbers representing the impulse response of the system. In our case, $\Omega_1 = -\omega_s/2$ and $\Omega_2 = \omega_s/2$, so $\Omega_2 - \Omega_1 = \omega_s$ and $2\pi/\omega_s = T$. Therefore, the pair of equations 3.7 and 3.8 can be rewritten as

$$H(\omega) = \sum_{n=-\infty}^{\infty} h(n) e^{-jn\omega T} \tag{3.9}$$

$$h(n) = \frac{1}{\omega_s} \int_{-\omega_s/2}^{\omega_s/2} H(\omega) e^{jn\omega T} \, d\omega \tag{3.10}$$

where we have substituted equation 3.3 for a_n.

We note that equation 3.9 is the same as equation 3.1 for $z = e^{j\omega T}$, assuming $h(n) = 0$ for $n < 0$, and having a finite upper limit N. These points are clarified in the design discussed in section 3.2.

Another important property of nonrecursive filters is obtained from the transfer function. Setting $z = e^{j\omega T}$ in equation 3.4, we have

$$H(\omega) = \sum_{n=0}^{2N} h(n)e^{-jn\omega T} = h(0) + h(1)e^{-j\omega T} + h(2)e^{-j2\omega T} + \ldots + h(N)e^{-jN\omega T} \ldots$$

$$+ h(2N-2)\,e^{-j(2N-2)\omega T} + h(2N-1)\,e^{-j(2N-1)\omega T}$$

$$+ h(2N)\,e^{-j2N\omega T}$$

where in order to simplify we have taken the upper limit as $2N$. Extracting the middle term factor

$$H(\omega) = e^{-jN\omega T}[h(0)\,e^{jN\omega T} + h(1)\,e^{j(N-1)\omega T} + h(2)\,e^{j(N-2)\omega T} + \ldots + h(N)$$

$$+ \ldots h(2N-2)\,e^{-j(N-2)\omega T} + h(2N-1)\,e^{-j(N-1)\omega T} + h(2N)\,e^{-jN\omega T}]$$

The function within the square bracket can become a real function of ωT for two special cases:

(i) $h(0) = h(2N)$, $h(1) = h(2N-1)$, $h(2) = h(2N-2)$, etc. (3.11)
(ii) $h(0) = -h(2N)$, $h(1) = -h(2N-1)$, $h(2) = -h(2N-2)\ldots$, $h(N) = 0$ (3.12)

In the first case (symmetrical impulse response), we have

$$H(\omega) = e^{-jN\omega T}[2\sum_{n=0}^{N-1} h(n)\cos(N-n)\omega T] \tag{3.13}$$

and in the second case (antisymmetrical impulse response), we have

$$H(\omega) = je^{-jN\omega T}[2\sum_{n=0}^{N-1} h(n)\sin(N-n)\omega T] \tag{3.14}$$

In both cases the phase response (θ) is a linear function of frequency. In the first case $\theta = -N\omega T$, and in the second case $\theta = \pi/2 - N\omega T$. Furthermore, they are independent of the coefficients $h(n)$. The amplitude response is a cosine and sine function for the first and second case respectively, and it can be adjusted by choice of coefficients $h(n)$. These arrangements have also been used in the past with the transversal filter networks described by Linke(16).

3.2 Design procedure

The periodic frequency response $H(\omega)$ and the corresponding impulse sequence $h(n)$ are linked by the Fourier pair of equations 3.9 and 3.10. The advantage of the Fourier series application, in this case, is that for a given $H(\omega)$ we immediately obtain the impulse response $h(n)$ by solving equation 3.10, and therefore the coefficients for the filter $a_n = h(n)$. In this procedure it is usually assumed that $H(\omega)$ has an idealized frequency response, as shown in fig. 3.1, which greatly simplifies solving the integral in equation 3.10. The solution for $h(n)$ contains an infinite number of terms due to the infinitely sharp (discontinuous) cutoff of the ideal filter characteristic. To make the solutions practically realizable $h(n)$ must be truncated to a finite length sequence. This in turn causes a spread of the frequency response into the stopband, and to control this we introduce a suitable window function to modify the $h(n)$

sequence. These points will be illustrated below on an ideal lowpass filter, and also in section 3.3 where we introduce window functions. The transformations from lowpass to other filter types are also discussed in this section.

Example 3.1
Consider the ideal lowpass frequency response characteristic with cutoff frequency ω_c within the range $\pm\omega_s/2$ as shown in fig. 3.2. Applying equation 3.10, we have

$$h(n) = \frac{1}{\omega_c}\int_{-\omega_s/2}^{\omega_s/2} H(\omega)e^{jn\omega T}\,d\omega = \frac{1}{\omega_s}\int_{-\omega_c}^{\omega_c} e^{j(n-\lambda)\omega T}\,d\omega$$

with the solution

$$h(n) = \frac{\omega_c T}{\pi}\,\frac{\sin(n-\lambda)\,\omega_c T}{(n-\lambda)\omega_c T} \tag{3.15}$$

where $n = \lambda, \lambda \pm 1, \lambda \pm 2, \ldots$. This is an infinite sequence of type $(\sin x)/x$ centred at λ, which represents the filter delay.

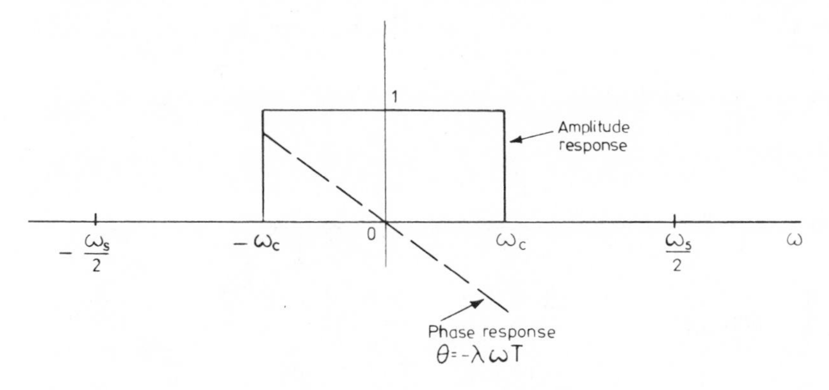

Fig. 3.2 Ideal lowpass filter frequency response

Example 3.2
To simplify the analysis, the phase characteristic is often neglected by assuming zero delay, i.e. $\lambda = 0$. We follow the same practice but introduce λ again at a later stage.

For the cutoff $\omega_c = \omega_s/8$, and $\lambda = 0$, equation 3.15 gives us the following impulse response:

$$h(n) = \tfrac{1}{4}\left(\frac{\sin n\pi/4}{n\pi/4}\right) \tag{3.16}$$

where $n = 0, \pm 1, \pm 2, \ldots, \pm\infty$. This is an infinite series of terms which cannot be realized because it would require an infinite number of filter coefficients and an infinite delay. To make it realizable we truncate the series to 21 terms as shown in fig. 3.3(a). Introducing now a finite delay of ten units of time (i.e. $\lambda = 10$), we obtain $h(n-10)$ series with values given in the first column of Table 3.1, and also shown in fig. 3.3(b). This is now a realizable (causal) impulse response since it is zero for negative values of n. The values in the second and third columns will be referred to in the next section.

Because the impulse response is symmetrical around $\lambda = N = 10$, we can calculate its frequency amplitude response from equation 3.13. Expanding this equation and using the

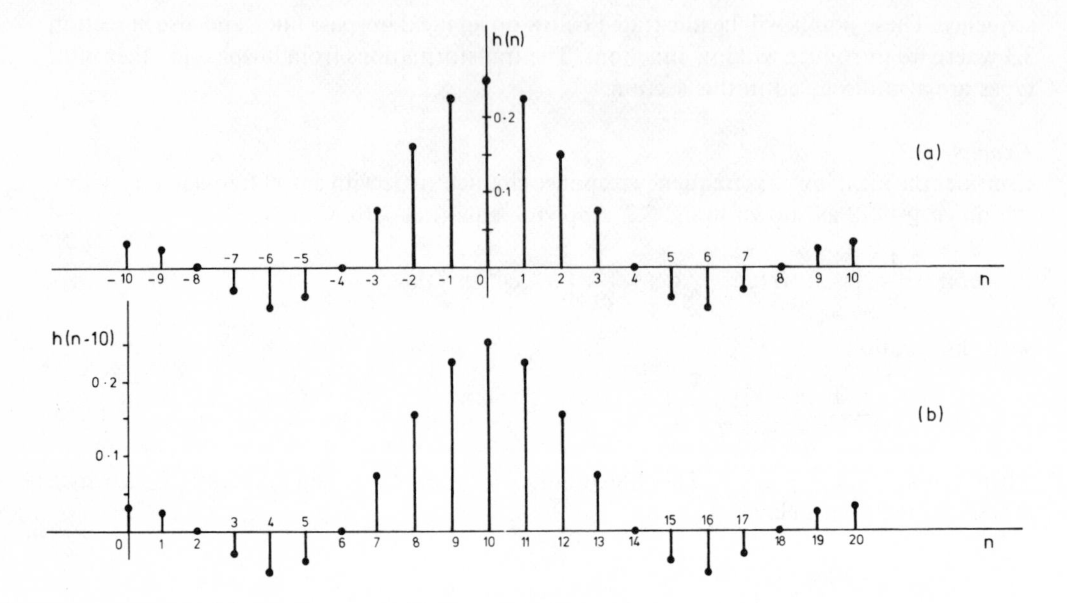

Fig. 3.3 (a) Truncated impulse response for ideal lowpass filter with $\omega_c = \omega_s/8$, (b) as (a) but with finite delay

Table 3.1

$h(n-10)$	R	H	K
$h(0) = h(20)$	0·03183099	0·00254648	0·00036542
$h(1) = h(19)$	0·02500879	0·00256375	0·00113711
$h(2) = h(18)$	0	0	0
$h(3) = h(17)$	−0·03215415	−0·00866936	−0·00637989
$h(4) = h(16)$	−0·05305165	−0·02110671	−0·01698820
$h(5) = h(15)$	−0·04501582	−0·02430854	−0·02092616
$h(6) = h(14)$	0	0	0
$h(7) = h(13)$	0·07502636	0·06079995	0·05758099
$h(8) = h(12)$	0·15915494	0·14517283	0·14168995
$h(9) = h(11)$	0·22507908	0·22001165	0·21867624
$h(10)$	0·25	0·25	0·25

Column R: truncated impulse response series (rectangular window)
Column H = Column R × Hamming window weights
Column K = Column R × Kaiser window weights

values from Table 3.1, we obtain the following expression for the amplitude response:

$$\begin{aligned} |H(m)|_R = 0.25 &+ 0.45015816 \cos m\pi \\ &+ 0.31830988 \cos 2m\pi \\ &+ 0.15005272 \cos 3m\pi \\ &- 0.09003164 \cos 5m\pi \\ &- 0.10610330 \cos 6m\pi \\ &- 0.06430830 \cos 7m\pi \\ &+ 0.05001758 \cos 9m\pi \\ &+ 0.06366198 \cos 10m\pi \end{aligned} \tag{3.17}$$

where we have introduced $\omega T = m\pi$ with $0 \leq m \leq 1$. In fact $m = \omega/(\omega_s/2)$ is the normalized frequency with respect to half the sampling frequency. The values of coefficients in Table 3.1 have been calculated using an electronic pocket calculator, but the amplitude response $|H(m)|$ has been calculated using the computer program given in section 4 of the appendix. The result is shown in fig. 3.4 by the graph marked as 21 coefficients. To illustrate the effect of the number of coefficients on $|H(m)|$ response, we have performed calculations for 11 and 31 coefficients and these too are marked in fig. 3.4.

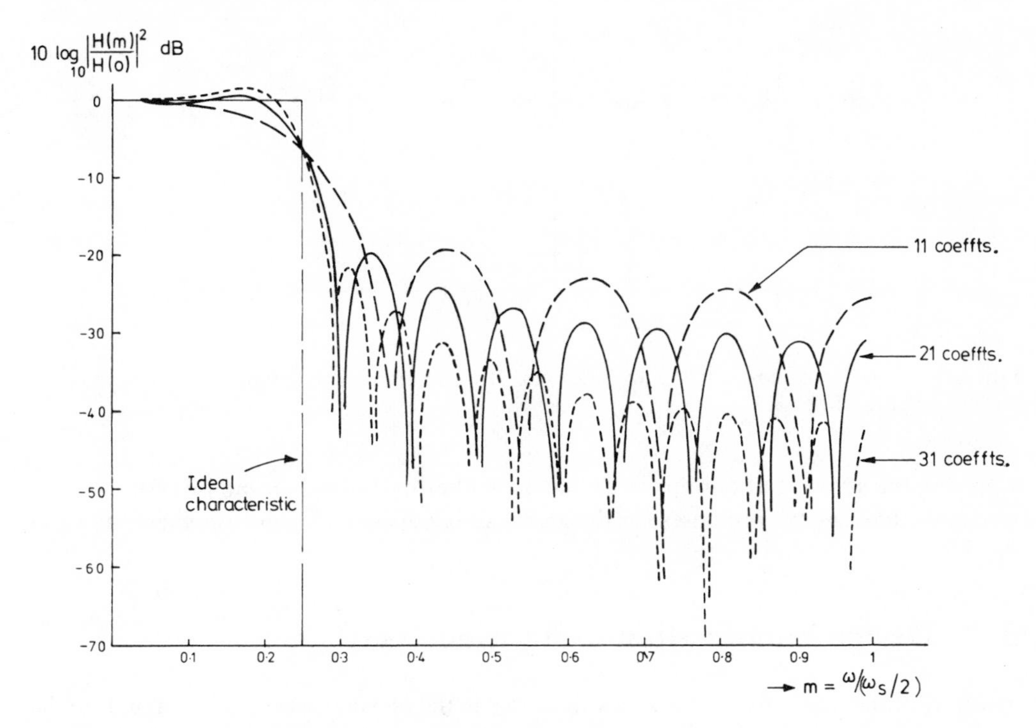

Fig. 3.4 Amplitude response for various filter lengths

Before proceeding to the next section, in which window functions are introduced, we briefly discuss the relationships between impulse responses of different types of filters. It can be shown that the unit-sample or impulse response of the highpass filter with the amplitude response shown in fig. 3.1(b), with $\omega_c = (\omega_s/2) - (\omega_c)_{LP}$, is given by

$$h(n)_{HP} = (-1)^n h(n)_{LP} \qquad \text{for } n = 0, \pm 1, \pm 2, \ldots \tag{3.18}$$

where $h(n)_{LP}$ is the lowpass unit-sample response.

For the bandpass filter of fig. 3.1(c), we have

$$h(n)_{BP} = (2\cos n\omega_0 T)\, h(n)_{LP} \qquad \text{for } n = 0, \pm 1, \pm 2, \ldots \tag{3.19}$$

where ω_0 is the bandpass centre frequency, and $\omega_1 = \omega_0 - \omega_c$, $\omega_2 = \omega_0 + \omega_c$, where ω_c refers to the lowpass filter as in fig. 3.1(a). The unit-sample response of the bandstop filter in fig. 3.1(d) is related to the bandpass filter in the following way:

$$\begin{aligned} h(0)_{BS} &= 1 - h(0)_{BP} \\ h(n)_{BS} &= -h(n)_{BP} \qquad \text{for } n = \pm 1, \pm 2, \ldots \end{aligned} \tag{3.20}$$

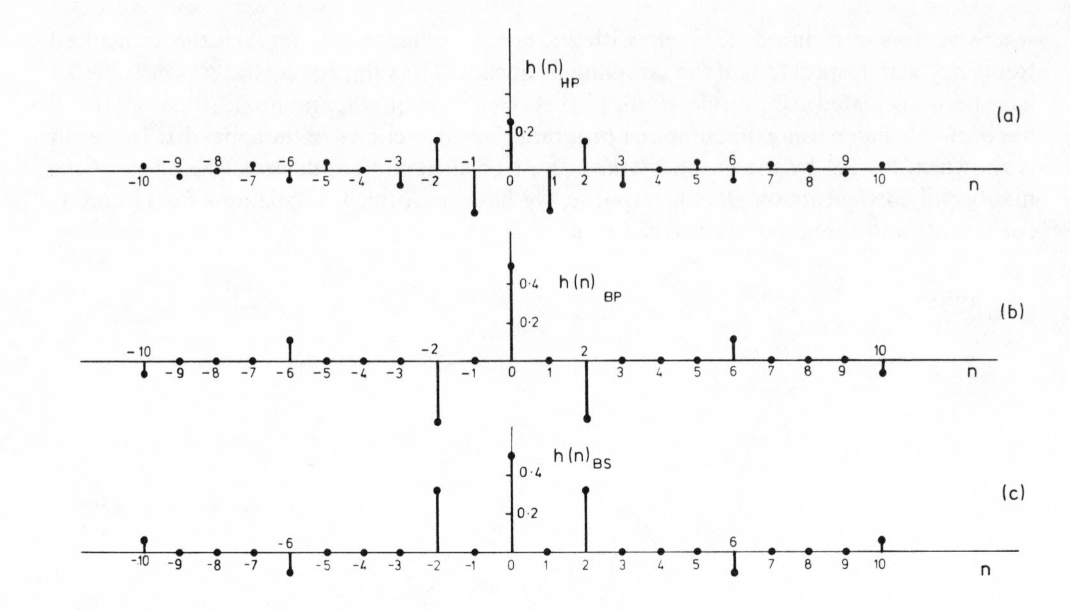

Fig. 3.5 Unit-sample response for (a) highpass, (b) bandpass and (c) bandstop filters based on the lowpass of fig. 3.3

Applying the above relationships to the lowpass filter unit-sample response of fig. 3.3, we obtain the unit-sample responses for the highpass, bandpass and bandstop filters shown in fig. 3.5.

3.3 Design modification using windows

The truncation of an infinite time series, discussed in the previous section, corresponds to the application of a rectangular window function defined as

$$w_R(n) = \begin{cases} 1 & |n| \leq N \\ 0 & |n| > N \end{cases}$$

The frequency response of a truncated time series can be improved by using a number of window functions which modify the unit-sample response $h(n)$ in a prescribed way by multiplication, i.e. $h(n) \times w(n)$. We discuss briefly two typical cases known as the generalized Hamming and Kaiser window functions. The generalized Hamming window function is given by

$$w_H(n) = \begin{cases} \alpha + (1-\alpha)\cos(n\pi/N) & |n| < N \\ 0 & |n| > N \end{cases} \tag{3.21}$$

where $0 \leq \alpha \leq 1$. If $\alpha = 0.54$ the window is called a Hamming window, and if $\alpha = 0.50$ it is called a Hanning window. An application of the Hamming window is illustrated in the example shown below.

Example 3.3
This is an extension of example 3.2. Choosing $\alpha = 0.54$, and $N = 10$ (from example 3.2),

equation 3.21 becomes

$$w_H(n) = 0.54 + 0.46 \cos(n\pi/10)$$

This function is symmetrical around the value of $w_H(0)$ which corresponds to the central point of the rectangular window in fig. 3.3(a). Using a pocket calculator we easily obtain the following values:

$w_H(0) = 1$
$w_H(1) = 0.97748600$
$w_H(2) = 0.91214782$
$w_H(3) = 0.81038122$
$w_H(4) = 0.678$
$w_H(5) = 0.54$
$w_H(6) = 0.39785218$
$w_H(7) = 0.26961878$
$w_H(8) = 0.172$
$w_H(9) = 0.10251400$
$w_H(10) = 0.08$

Multiplying the values of column R in Table 3.1 by the above weights we obtain column H. Note that $w_H(0)$ corresponds to $h(10)$ in column R of Table 3.1.

Using the values of column H, in equation 3.13, we obtain the expression for the amplitude response with Hamming window, as

$$\begin{aligned} |H(m)|_H = 0.25 &+ 0.44002330 \cos m\pi \\ &+ 0.29034566 \cos 2m\pi \\ &+ 0.12159990 \cos 3m\pi \\ &- 0.04861708 \cos 5m\pi \\ &- 0.04221342 \cos 6m\pi \\ &- 0.01733872 \cos 7m\pi \\ &+ 0.00512750 \cos 9m\pi \\ &+ 0.00509296 \cos 10m\pi \end{aligned} \tag{3.22}$$

with the same comments as given for equation 3.17. The results of computer calculations as in section 4 of the appendix for the ratio

$$20 \log_{10} |H(m)/H(0)| \text{ (dB)}$$

where $H(m)$ is given by equation 3.22, are shown in fig. 3.6 under the Hamming window, where the result for the rectangular window of example 3.2 is also shown for comparison.

The second window we discuss is a family of weighting functions proposed by Kaiser (9),

$$w_K(n) = \begin{cases} \dfrac{I_0[\beta\sqrt{(1-(n/N)^2)}]}{I_0(\beta)} & |n| \leq N \\ 0 & |n| > N \end{cases} \tag{3.23}$$

where I_0 is the modified Bessel function of the first kind and zero order. The parameter β specifies its frequency response in terms of the main lobe width and the side lobe level. Large values of β correspond to a wider main lobe width and smaller side lobe levels. The normal range of β is $4 \leq \beta \leq 9$, corresponding to a range of side lobe peak heights of 3.1 % down to 0.047 %.

The weighting function given in equation 3.23 is symmetrical around $w_K(0)$, so it is sufficient to calculate its values for $n = 0, 1, \ldots, N$. These values are then used to modify the $h(n)$ response.

Example 3.4
This is again an extension of example 3.2, but now using the Kaiser window. Choosing the value $\beta = 2\pi$ in equation 3.23, and $N = 10$ as in example 3.2, we have

$$w_K(n) = \frac{I_0[2\pi\sqrt{(1-(n/10)^2)}]}{I_0(2\pi)}$$

where $n = 0, 1, 2, \ldots, 10$. The difficulty is that the tabulated Bessel functions in mathematical handbooks do not give $I_0(x)$ for all values of x, so employing the power series expansion

$$I_0(x) = 1 + \sum_{m=1}^{M} \left(\frac{(x/2)^m}{m!}\right)^2 \tag{3.24}$$

and evaluating this expression to within specified accuracy using the computer program proposed by Kaiser and given in section 4 of the appendix (also in Rabiner & Gold (15), p. 103), we obtain the values for the weights given in Table 3.2. The last column of Table 3.2 gives the upper limit M used in the sum of equation 3.24, based on the criterion that the last term in the series is less than 1×10^{-8}.

Table 3.2 Weights for nonrecursive filter using Kaiser window function

n	$w_K(n)$	M
0	1·0000000000	15
1	0·9715529534	14
2	0·8902642058	14
3	0·7674768355	14
4	0·6195788531	14
5	0·4648623206	13
6	0·3202200325	13
7	0·1984156504	12
8	0·1064793789	11
9	0·0454683321	9
10	0·0114799346	1

Multiplying the values of column R in Table 3.1 by the weights given in Table 3.2 we obtain the column K. We note again that $w_K(0)$ corresponds to $h(10)$ in column R of Table 3.1. Using the values of column K in equation 3.13 we obtain the following expression for the amplitude response.

$$\begin{aligned} |H(m)|_K = 0.25 &+ 0.43735248 \cos m\pi \\ &+ 0.28337990 \cos 2m\pi \\ &+ 0.11516198 \cos 3m\pi \\ &- 0.04185232 \cos 5m\pi \\ &- 0.03397640 \cos 6m\pi \\ &- 0.01275978 \cos 7m\pi \\ &+ 0.00227422 \cos 9m\pi \\ &+ 0.00073084 \cos 10\,m\pi \end{aligned} \tag{3.25}$$

with explanations as given for equation 3.17. The response has been calculated as

$$20 \log_{10} |H(m)/H(0)| \text{ (dB)}$$

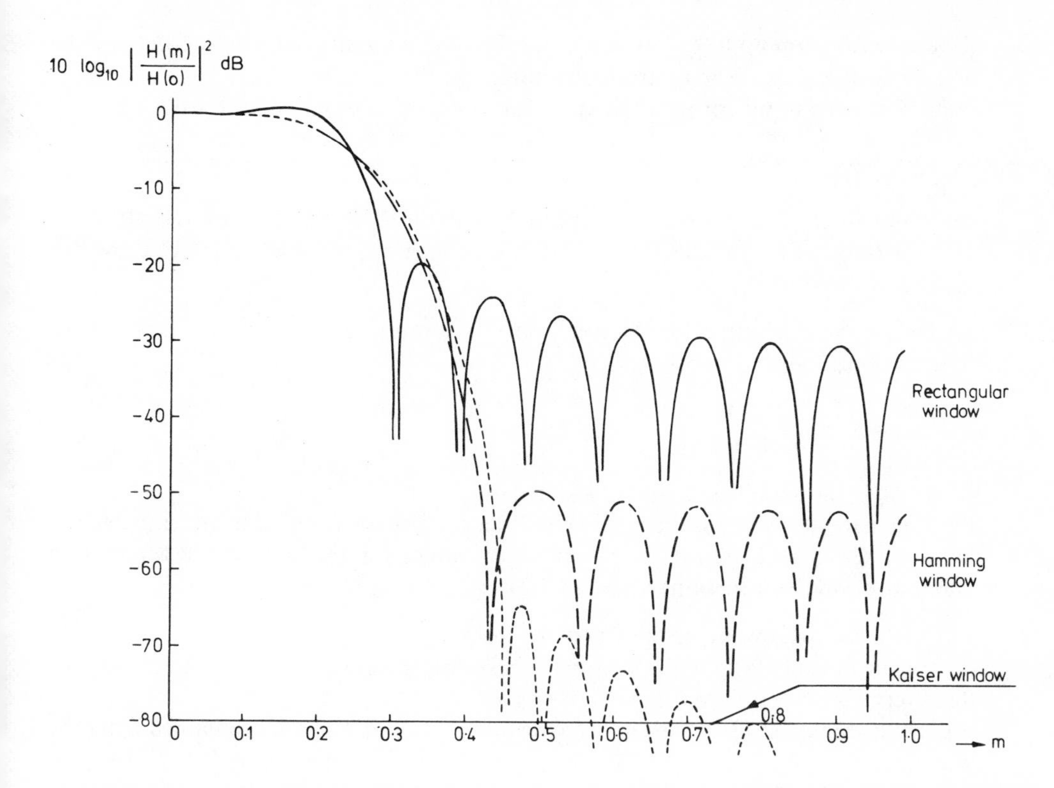

Fig. 3.6 Amplitude response using rectangular window, Hamming window and Kaiser window functions

using the computer program of section 4 of the appendix, and shown in fig. 3.6 under the Kaiser window.

Comparing fig. 3.4 with fig. 3.6 we see that the window shaping of filter coefficients is more powerful than the extension of the number of coefficients from 21 to 31 in reducing the stopband response. If the number of coefficients was increased considerably, this would produce nearly ideal frequency response, but such a solution would be unacceptable in practice, mainly on the grounds of cost.

The multiplication of the impulse response by the window function in the time-domain, $h(n) \times w(n)$, corresponds to the convolution in the frequency-domain, i.e. $H(\omega) * W(\omega)$. The window weighting function $w(t)$ is chosen so that $W(\omega)$ is narrowband. The major effect of the window is that discontinuities in $H(\omega)$ become transition bands on either side of discontinuity. Frequency responses of various window functions can be found, for example, in Oppenheim & Schaffer(10), section 5.5, and Rabiner & Gold(15), section 3.12.

3.4 Problems

3.1 The analogue differentiator is described by the transfer function

$$H(s) = sT$$

where s is the Laplace variable. Design a nonrecursive digital differentiator to simulate

the above function up to $f_s/2$, where $f_s = 1/T$ is the sampling frequency. Calculate the first 19 coefficients of the nonrecursive filter.
Note: For solving this problem, it is useful to know $\int x\,e^{jx}\,dx = e^{jx}(1 - jx)$.

3.2 (a) Calculate the magnitude of the frequency response for three and five middle terms in the solution of problem 3.1.
(b) Compare the above two cases graphically with the ideal differentiator $|H(\omega)| = \omega T$ over the range of frequencies 0 to $\omega_s/2$. Observe the improvement made by taking five instead of three terms.

3.3 (a) Derive the expression for the coefficients of a nonrecursive filter which is to have frequency response

$$H(\omega) = \begin{cases} 1 & 0 \leq \omega \leq \omega_s/4 \\ 0 & \text{elsewhere} \end{cases}$$

where ω_s is the sampling frequency in rad s^{-1}.
(b) Calculate the values of the first twelve filter coefficients from the expression derived in problem 3.3(a). Determine their modified values for the case of the generalized Hamming window function, equation 3.21.

3.4 Calculate the magnitude of the frequency response:
(a) for the lowpass filter with the impulse response given in fig. 3.3(a), taking only the three terms $h(0) = 0.25$ and $h(1) = h(-1) = 0.23$;
(b) for the highpass filter with impulse response given in fig. 3.5(a) obtained from the lowpass case using equation 3.18, taking only the three terms $h(0) = 0.25$ and $h(1) = h(-1) = -0.23$. Plot and compare responses of (a) and (b) over the frequency range $-\omega_s/2$ to $\omega_s/2$.

3.5 Calculate the magnitude of the frequency response:
(a) for the bandpass filter with the impulse response given in fig. 3.5(b), obtained from the lowpass case using equation 3.19, taking only the first three nonzero terms $h(0) = 0.50$ and $h(-2) = h(2) = -0.32$;
(b) for the bandstop filter with the impulse response given in fig. 3.5(c), obtained from the above bandpass case using equation 3.20, taking only the first three nonzero terms $h(0) = 0.50$ and $h(-2) = h(2) = 0.32$.
Plot and compare responses of (a) and (b) over the frequency range $-\omega_s/2$ to $\omega_s/2$. Compare these graphs with the ones of problem 3.4.

4

Recursive filter design

4.0 Introduction

The division of digital filters into nonrecursive and recursive types has been discussed earlier, particularly in section 2.5. The recursive filter structure and the form of its transfer function dictate design approaches to be different from those used in nonrecursive filter design. Rewriting equation 2.4 gives the general recursive filter transfer function as

$$H(z) = \frac{\sum_{n=0}^{N} a_n z^{-n}}{1 + \sum_{m=1}^{M} b_m z^{-m}} \tag{4.1}$$

The problem of recursive filter design is to determine the filter coefficients, a_n and b_m, such that the filter specifications are satisfied.

There are two basic approaches to this design problem. The first, a direct approach, is to determine the coefficients of the digital filter by some computational procedure directly from the filter specifications, and is described by Bogner & Constantinides (17), chapter 5. The second basic approach, to be considered here, is to determine the coefficients in an indirect way from the analogue (i.e. continuous-time) filters. This method consists of two parts:

(i) determination of a suitable analogue filter transfer function $H(s)$ which meets the required filter specification;
(ii) digitalization of this analogue filter.

The indirect method is implicit in the digital simulation of analogue filters: in this case the analogue filter is already known and only digitalization is required. Regarding design approach based on analogue filters, it is perhaps worthwhile to point out also that the recursive filter is a natural counterpart of the analogue filter as seen in section 1.1.

Digital filter characterization is specified in the frequency range $0 \leq \omega \leq \omega_s/2$, where ω_s is the sampling frequency, as discussed in section 3.0 and illustrated in fig. 3.1. We consider two main methods of recursive filter design, known as the impulse invariant method and the bilinear z-transform, presented in sections 4.1 and 4.2 respectively. These design techniques, developed for the lowpass filter, are extended to other types of filters by means of the frequency transformations discussed in section 4.3. Some general comments on accuracy concerning both the nonrecursive and recursive filters are given in section 4.4. Various examples are given throughout sections of this chapter, and additional information and exercises are given in section 4.5.

4.1 The impulse invariance method

As pointed out in the introduction, we start with the analogue filter transfer function $H(s)$ whose general form is given by

$$H(s) = \frac{\sum_{n=0}^{N} c_n s^n}{\sum_{n=0}^{M} d_n s^n} \tag{4.2}$$

If $M > N$, i.e. if the degree of the polynomial in the denominator is larger than the degree of the numerator, and if the poles of $H(s)$ are simple, then its partial-fractions form can be written as

$$H(s) = \sum_{n=1}^{M} \frac{A_n}{s + s_n} \tag{4.3}$$

where A_n are real or complex constants.

It is sufficient for our analysis to consider a typical simple pole term at $s = s_1$, for which we have

$$H_1(s) = \frac{A_1}{s + s_1} \tag{4.4}$$

The corresponding impulse response is

$$h_1(t) = A_1 \mathrm{e}^{-s_1 t} \tag{4.5}$$

and its sampled form is written directly as

$$h_1(kT) = A_1 \mathrm{e}^{-s_1 kT} \tag{4.6}$$

where T is the sampling interval and k is an integer. The sampled impulse frequency response is given by

$$H_{1s}(s) = T \sum_{k=0}^{\infty} h_1(kT) \mathrm{e}^{-skT} \tag{4.7}$$

where s is the complex frequency, and substituting equation 4.6 for $h_1(kT)$, we obtain an infinite geometric series whose sum produces the result

$$H_{1s}(s) = \frac{A_1 T}{1 - \mathrm{e}^{-s_1 T} \mathrm{e}^{-sT}} \tag{4.8}$$

Applying now the standard z-transformation $z = \mathrm{e}^{sT}$ from equation 1.19, we obtain the discrete-time (or sampled-data) transfer function

$$H_1(z) = \frac{A_1 T}{1 - \mathrm{e}^{-s_1 T} z^{-1}} \tag{4.9}$$

The impulse invariant method is also known as the standard z-transform method because of the application of $z = \mathrm{e}^{sT}$. We already know, from sections 1.4 and section 3 of the appendix, that the discrete-time transfer function 4.8 or 4.9 is a periodic function with period $\omega_s = 2\pi/T$.

The result equation 4.9 as a single term of equation 4.3 can be extended to n terms, in which case we have the following correspondence:

$$H(s) = \sum_{n=1}^{M} \frac{A_n}{s+s_n} \rightarrow \sum_{n=1}^{M} \frac{A_n T}{1-e^{-s_n T}z^{-1}} = H(z) \tag{4.10}$$

The simple one-pole analogue filter is transformed into a digital filter in the following way:

$$\frac{a}{s+a} \rightarrow \frac{aT}{1-e^{-aT}z^{-1}} \tag{4.11}$$

and for second-order systems

$$\frac{s+a}{(s+a)^2+b^2} \rightarrow T\frac{1-e^{-aT}(\cos bT)z^{-1}}{1-2e^{-aT}(\cos bT)z^{-1}+e^{-2aT}z^{-2}} \tag{4.12}$$

$$\frac{b}{(s+a)^2+b^2} \rightarrow T\frac{e^{-aT}(\sin bT)z^{-1}}{1-2e^{-aT}(\cos bT)z^{-1}+e^{-2aT}z^{-2}} \tag{4.13}$$

The last two relationships are obtained by expanding the left-hand-side into partial-fractions and then applying equation 4.11 to each individual term. Combining the two terms we obtain the right-hand-side of equations 4.12 and 4.13.

Example 4.1
We want to design a discrete-time filter of recursive type with the frequency response of an analogue single-section RC filter whose transfer function is given by

$$H(s) = \frac{\omega_c}{s+\omega_c} \tag{4.14}$$

where $\omega_c = 1/RC$ defines the -3dB cutoff point.

Applying the transformation of equation 4.11 to equation 4.14, we obtain the digital transfer function

$$H(z) = \frac{\omega_c T}{1-z^{-1}e^{-\omega_c T}} \tag{4.15}$$

The discrete-time filter structure is obtained from the above equation by expressing $H = Y/X$, where Y and X refer to the filter output and input respectively (see, for example, section 2.4). It is easy to show that the filter structure is given by the difference equation

$$y(k) = (\omega_c T)x(k) + e^{-\omega_c T}y(k-1) \tag{4.16}$$

which is shown in fig. 4.1 (see also fig. 1.2(a)).

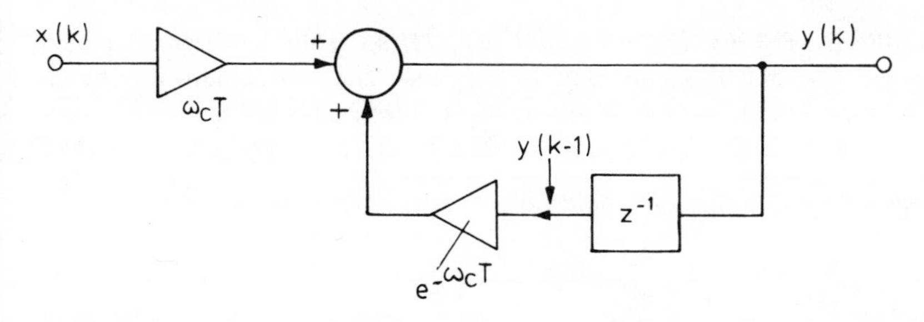

Fig. 4.1 Recursive digital structure equivalent to analogue single-section RC filter

To compare the frequency responses of the above filters we have chosen $\omega_c = 2\pi \times 10^3 \text{rad s}^{-1}$, $\omega_s = 2\pi \times 10^4 \text{rad s}^{-1}$, for which $\omega_c T = 0.628$. With these values, the analogue filter magnitude squared has been arranged in the form

$$|H(n)|^2 = \frac{1}{1 + n^2/16} \tag{4.17}$$

obtained by the choice of the frequency steps $\Delta f = 250\,\text{Hz}$. The function $10 \log_{10} |H(n)/H(0)|^2$ has been calculated, using a computer, for $n = 0, 1, 2, 3, \ldots, 40$ which covers the frequency range up to $40\Delta f = 10\,\text{kHz}$. The result is shown in fig. 4.2, graph (a). The result is shown in fig. 4.2, graph (a).

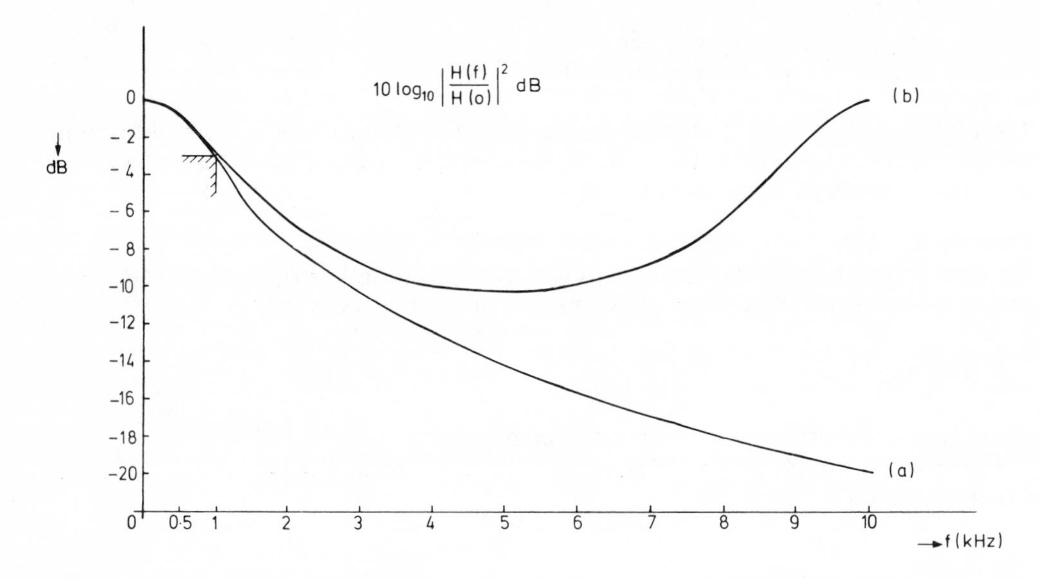

Fig. 4.2 Frequency response for (a) analogue and (b) digital version of RC lowpass filter

With $\omega_c T = 0.628$, the digital filter magnitude squared response for equation 4.15 can be arranged as

$$|H(m)|^2 = \frac{0.394}{1.285 - 1.068 \cos m\pi} \tag{4.18}$$

where we have introduced $\omega T = m\pi$. The function $10 \log_{10} |H(m)/H(0)|^2$ has been calculated from $m = 0$ to $m = 2$ in steps of 0.05, section 5 of the appendix. The result obtained is shown in fig. 4.2, graph (b), for a comparison with the analogue filter graph (a).

Example 4.2
We determine a digital filter for the three-pole Butterworth lowpass filter

$$H(s) = \frac{s_1 s_2 s_3}{(s + s_1)(s + s_2)(s + s_3)} \tag{4.19}$$

where $s_1 = \omega_c$, $s_2 = \frac{1}{2}(1 + j\sqrt{3})\omega_c$, $s_3 = \frac{1}{2}(1 - j\sqrt{3})\omega_c$; ω_c is the cutoff frequency defined by

$|H(\omega_c)| = 0.707$, as explained by Gold & Rader(12), p. 57. Expanding equation 4.19 into partial-fractions we have

$$H(s) = \frac{A_1}{s+s_1} + \frac{A_2}{s+s_2} + \frac{A_3}{s+s_3} \tag{4.20}$$

with $A_1 = \omega_c$, $A_2 = -\frac{1}{2}(1 - \mathrm{j}.1/\sqrt{3})\omega_c$, and $A_3 = -\frac{1}{2}(1 + \mathrm{j}.1/\sqrt{3})\omega_c$.

Applying the transformation given by equation 4.11 to each term in equation 4.20 we obtain the following result:

$$H(z) = \frac{\omega_c T}{1 - \mathrm{e}^{-\omega_c T} z^{-1}} + \frac{[-1 + f(\omega_c T) z^{-1}]\omega_c T}{1 - [2\mathrm{e}^{-\omega_c T/2} \cos(\sqrt{\tfrac{3}{2}}\omega_c T)] z^{-1} + \mathrm{e}^{-\omega_c T} z^{-2}} \tag{4.21}$$

where $f(\omega_c T) = \mathrm{e}^{-\omega_c T/2}[\cos(\sqrt{\tfrac{3}{2}}\omega_c T) + \sqrt{\tfrac{1}{3}} \sin(\sqrt{\tfrac{3}{2}}\omega_c T)]$

This result is in the form suitable for the parallel realization of fig. 2.8.

To derive the difference equation, and hence the filter structure, we express $H(z)$ as

$$H(z) = \frac{Y(z)}{X(z)} = \frac{Y_1(z)}{X(z)} + \frac{Y_2(z)}{X(z)} \tag{4.22}$$

Comparing equations 4.21 and 4.22, we have

$$H_1(z) = \frac{Y_1(z)}{X(z)} = \frac{\omega_c T}{1 - \mathrm{e}^{-\omega_c T} z^{-1}}$$

from which we obtain the difference equation

$$y_1(k) = (\omega_c T) x(k) + (\mathrm{e}^{-\omega_c T}) y_1(k-1) \tag{4.23}$$

Similarly for the second term,

$$y_2(k) = (-\omega_c T) x(k) + [\omega_c T f(\omega_c T)] x(k-1) + [2\mathrm{e}^{-\omega_c T/2} \cos(\sqrt{\tfrac{3}{2}}\omega_c T)] y_2(k-1)$$
$$- (\mathrm{e}^{-\omega_c T}) y_2(k-2) \tag{4.24}$$

The complete solution is the sum of equations 4.23 and 4.24

$$y(k) = y_1(k) + y_2(k)$$

defining the digital filter structure shown in fig. 4.3.

We have also calculated magnitude responses for the analogue and digital filter transfer functions given by equations 4.19 and 4.21 respectively. The values $\omega_c = 2\pi \times 10^3$ rad s^{-1}, $\omega_s = 2\pi \times 10^4$ rad s^{-1} and $\omega_c T = 0.628$ have been chosen as in example 4.1.

For the analogue filter of equation 4.19, we have

$$|H(n)|^2 = \frac{1}{1 + n^6/4096} \tag{4.25}$$

for frequency steps of 250 Hz and $n = 0, 1, 2, 3, \ldots, 40$.

Similarly, for the digital filter of equation 4.21, we have

$$|H(m)|^2 = \frac{0.009 + 0.008 \cos m\pi}{5.712 - 8.546 \cos m\pi + 3.421 \cos 2m\pi - 0.570 \cos 3m\pi} \tag{4.26}$$

for $\omega T = m\pi$ and $0 \leq m \leq 2$, in steps of 0.05.

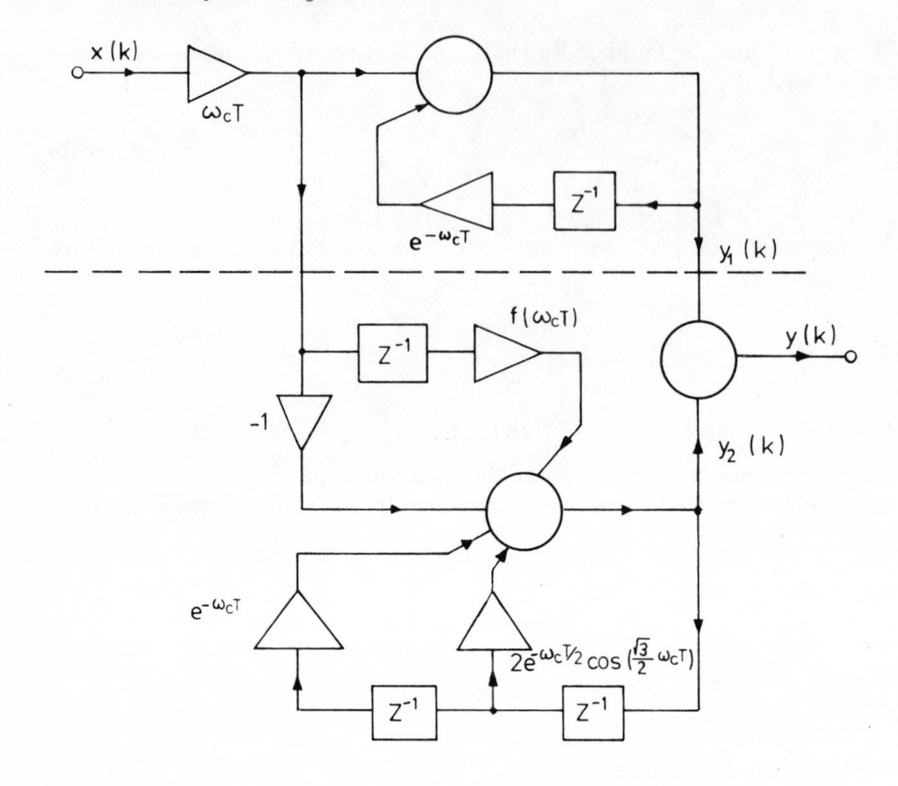

Fig. 4.3 Digital structure for three-pole Butterworth lowpass filter

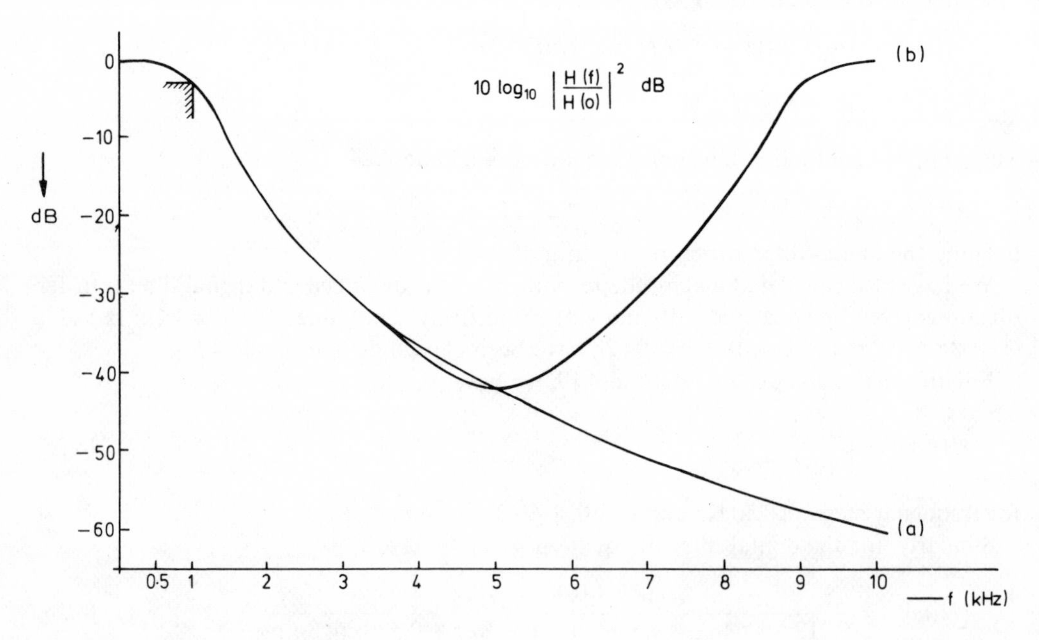

Fig. 4.4 Frequency response for (a) analogue and (b) digital versions of three-pole Butterworth lowpass filter

Using the above equations and the computer program given in section 5 of the appendix, we have calculated the quantity $10 \log_{10}|H/H(0)|^2$ for each case. Results are shown in fig. 4.4(a) for the analogue filter (equation 4.25), and in fig. 4.4(b) for the digital version of this filter (equation 4.26).

There are two problems inherent in the design of digital filters by the impulse invariant method. One is caused by the interspectra interference (spectrum 'folding') introduced by terms $H(\omega + n\omega_s)$ for $n \neq 0$. If the analogue filter transfer function $H(s)$ is bandlimited to the baseband, i.e. $H(\omega) = 0$ for $\omega > \omega_s/2$, then there is no folding error and the frequency response of the digital filter is identical to that of the original analogue filter. However, when $H(s)$ is a rational function of s, it is not bandlimited and therefore $H(s) \neq H_s(s)$ in the baseband. The magnitude of errors resulting from the folding is directly related to the high frequency asymptotic behaviour of $H(s)$, as seen by comparing graphs (a) and (b) in figs 4.2 and 4.4. To reduce the possibility of such errors, the function $H(s)$ can be modified by adding in cascade a wideband lowpass filter as mentioned in section 1.4.

The other problem is the variation of the filter gain with sampling frequency seen, for example, in equation 4.15 for $z = 1$, i.e. at $\omega = 0$. It may be necessary to compensate for it, in order to protect against the possibility of overflow in the computer program.

4.2 The bilinear *z*-transformation method

In the previous section we have dealt with one term in the partial-fraction expansion of an analogue filter given by equation 4.4, which can be rewritten as

$$H(s_a) = \frac{A_1}{s_a + s_1} \tag{4.27}$$

The variable s_a is used in place of s as a more definite notation for the analogue filter. We introduce now the transformation

$$s_a = \frac{2}{T} \tanh \frac{s_d T}{2} \tag{4.28}$$

where s_d is a new variable whose properties and effects are discussed below. Substituting equation 4.28 into 4.27, and setting $s_a = j\omega_a$ and $s_d = j\omega_d$ we obtain the frequency response

$$H(\omega_d) = \frac{A_1}{s_1 + j\dfrac{2}{T} \tan \dfrac{\omega_d T}{2}} \tag{4.29}$$

and the transformation, equation 4.28, becomes

$$\omega_a = \frac{2}{T} \tan \frac{\omega_d T}{2} \tag{4.30}$$

From the above expressions we can make the following observations. First we see that the analogue filter frequency response $H(\omega_a)$ in equation 4.27 tends to zero only as $\omega_a \to \infty$. However, the modified function $H(\omega_d)$ in equation 4.29 goes to zero for finite values of ω_d given by

$$\omega_d T/2 = (2\lambda + 1)\ \pi/2 \tag{4.31}$$

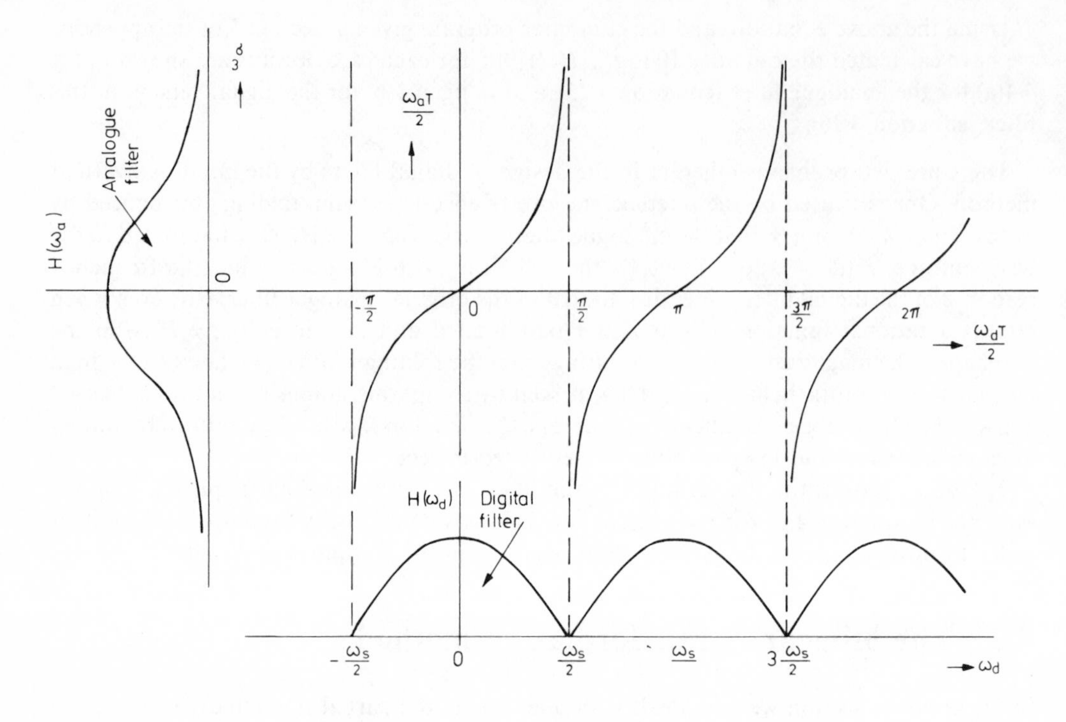

Fig. 4.5 Illustration of spectra compression due to the bilinear z-transformation

where $\lambda = 0, \pm 1, \pm 2, \ldots$. Therefore, we see that the transformation given in equation 4.28 compresses the infinite frequency range ω_a into a finite range, and makes the modified transfer function given by equation 4.29 periodic. These properties are illustrated in fig. 4.5, where the quantity T has been interpreted, as before, as the reciprocal of the sampling frequency ($T = 1/f_s$). This means that the spectra folding problem is eliminated since the base-band is confined to $\omega_s/2$, but the disadvantage to this is a distorted frequency scale as shown in fig. 4.6. As will be seen in later examples, this frequency scale distortion (or warping) is taken into account in the course of the digital filter design.

Returning to the transformation of equation 4.28, we can rewrite it as

$$s_a = \frac{2}{T}\frac{1 - e^{-s_d T}}{1 + e^{-s_d T}}$$

and applying the standard z-transformation, in this case $z = e^{s_d T}$, we obtain

$$s_a = \frac{2}{T}\frac{1 - z^{-1}}{1 + z^{-1}} \tag{4.32}$$

This relationship, known as the bilinear z-transformation, links the analogue filter variable s_a with the digital filter variable z. To obtain the digital filter transfer function $H(z)$ from a given analogue filter transfer function $H(s)$, we simply substitute from equation 4.32, giving

$$H(z) = H(s)\Big|_{s = \frac{2}{T}\frac{1 - z^{-1}}{1 + z^{-1}}} \tag{4.33}$$

This method of digital filter design is illustrated in the following examples.

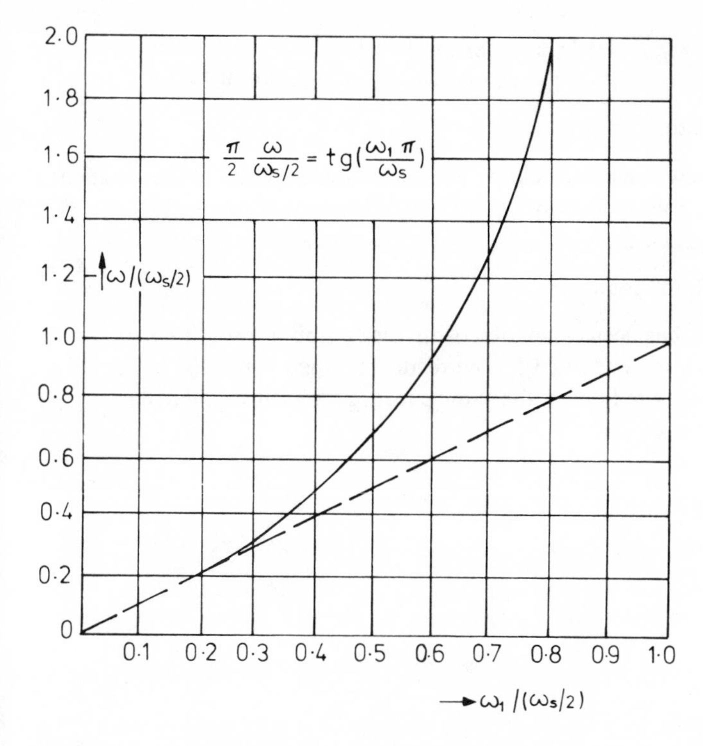

Fig. 4.6 Distorted frequency scale due to the bilinear z-transformation

Example 4.3

Consider the first-order analogue (RC) filter used in example 4.1, whose transfer function is

$$H(s_a) = \frac{\omega_{ac}}{\omega_{ac} + s_a} \tag{4.34}$$

where we have added the subscript a to denote the analogue filter. We want to design a digital first-order filter with cutoff frequency ω_{dc}. Then the equivalent analogue filter cutoff frequency is given by equation 4.30, i.e.

$$\omega_{ac} = \frac{2}{T}\tan\frac{\omega_{dc}T}{2} \tag{4.35}$$

The digital filter, as given by equation 4.33, is obtained from equation 4.34 with the application of equation 4.32

$$H(z) = \frac{(\omega_{ac}T/2)(1 + z^{-1})}{(\omega_{ac}T/2 + 1) + (\omega_{ac}T/2 - 1)z^{-1}} \tag{4.36}$$

From the above we obtain, as before, the difference equation

$$y(k) = \frac{\omega_{ac}T/2}{1 + \omega_{ac}T/2}[x(k) + x(k-1)] + \frac{1 - \omega_{ac}T/2}{1 + \omega_{ac}T/2}y(k-1) \tag{4.37}$$

which defines the hardware filter structure.

We calculate the frequency response of this filter by choosing $\omega_{dc} = 2\pi \times 10^3$ rad s^{-1} and

$\omega_s = 2\pi \times 10^4\, \text{rad s}^{-1}$, so that $\omega_{dc}T = 0.2\pi$ as in example 4.1.

The equivalent analogue filter cutoff is then obtained from equation 4.35, or

$$\omega_{ac}T/2 = \tan(\omega_{dc}T/2) = \tan(0.1\pi) = 0.325$$

Using this value in equation 4.36, we can calculate the magnitude squared of the frequency response and express it in the following way:

$$|H(m)|^2 = 0.212 \frac{1 + \cos m\pi}{2.212 - 1.789 \cos m\pi} \tag{4.38}$$

The quantity $10 \log_{10} |H/H(0)|$ has been calculated using the computer program in section 5 of the appendix, for $0 \leq m \leq 2$ in steps of 0.05. The result obtained is plotted in fig. 4.7 as graph (a), together with the analogue filter of the same cutoff, graph (b), calculated earlier in example 4.1.

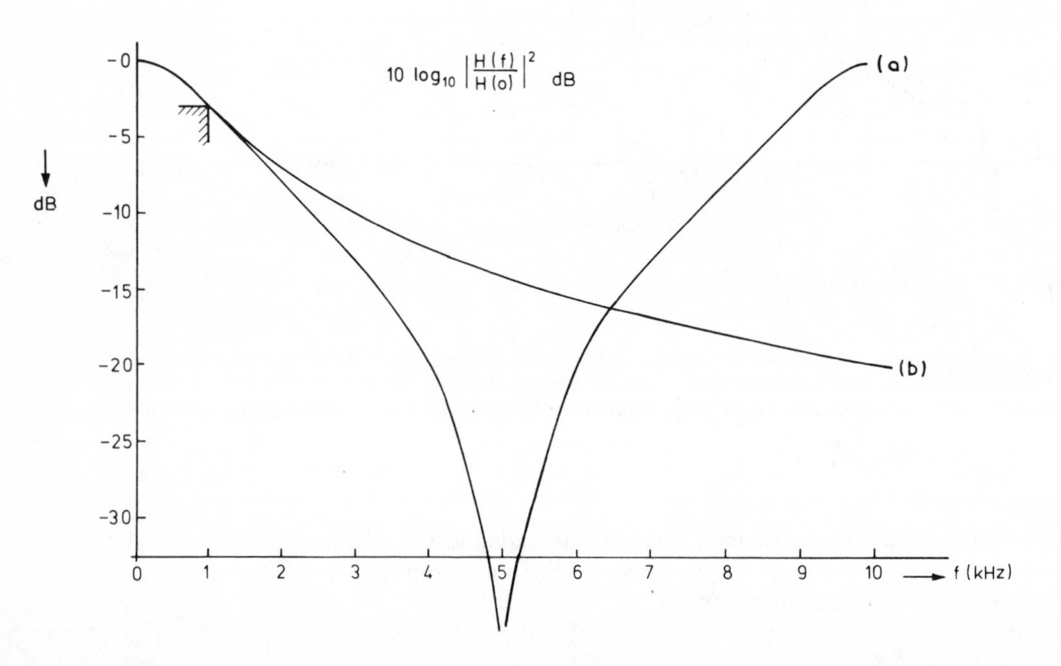

Fig. 4.7 Frequency response for first-order RC filter: (a) digital version with the bilinear z-transformation and (b) original analogue version

Example 4.4

We design a digital filter, for a 10 kHz sampling rate, which is flat to 3 dB in the passband of 0 to 1000 Hz, and which is more than 10 dB down at frequencies beyond 2000 Hz. The filter must be monotonic in passband and stopband. It is known from the analogue filter theory that a Butterworth filter can meet such a specification.

The specified characteristic frequencies are $\omega_{d1}T = 2\pi \times 10^3 \times 10^{-4} = 0.2\pi$, and $\omega_{d2}T = 0.4\pi$. The corresponding set of analogue frequencies are obtained by means of equation 4.30 as

$$\omega_{a1} = \tan(\omega_{d1}T/2) = \tan 0.1\pi = 0.325$$

$$\omega_{a2} = \tan(\omega_{d2}T/2) = \tan 0.2\pi = 0.726$$

The factor $2/T$ is not included because it cancels out within the ratio, seen for example in equations 4.39 and 4.40.

The Butterworth nth order filter (see, for example, Stanley (11)) is given by

$$|H(\omega)|^2 = \frac{1}{1+(\omega/\omega_c)^{2n}} \tag{4.39}$$

where $\omega_c = \omega_{a1} = 0.325$. The order n is found from the condition for the above response to be 10 dB down at $\omega_{a2} = 0.726$, i.e.

$$1 + (0.726/0.325)^{2n} = 10$$

giving $n = 1.432$, therefore we choose $n = 2$. A second-order Butterworth filter with $\omega_c = 0.325$ has poles at $s_{1,2} = 0.325\ (0.707 \pm \mathrm{j} \times 0.707) = 0.23 \pm \mathrm{j}0.23$, and no zeros. The transfer function is given by

$$H(s) = \frac{s_1 s_2}{(s+s_1)(s+s_2)} = \frac{0.1058}{s^2+0.46s+0.1058} \tag{4.40}$$

Replacing s by $(1-z^{-1})/(1+z^{-1})$ in the above expression for $H(s)$, we have

$$H(z) = \frac{0.1058}{\left(\dfrac{1-z^{-1}}{1+z^{-1}}\right)^2 + 0.46\left(\dfrac{1-z^{-1}}{1+z^{-1}}\right) + 0.1058}$$

or, after rearrangement

$$H(z) = 0.0676\frac{1+2z^{-1}+z^{-2}}{1-1.141z^{-1}+0.413z^{-2}} \tag{4.41}$$

which is the transfer function of the required digital filter. The corresponding difference equation is

$$\begin{aligned} y(k) = {} & 0.0676\ x(k) + 0.135\ x(k-1) + 0.0676\ x(k-2) \\ & + 1.141\ y(k-1) - 0.413\ y(k-2) \end{aligned} \tag{4.42}$$

which follows from equation 4.41.

The frequency response is given by

$$H(\omega) = H(z = \mathrm{e}^{\mathrm{j}\omega T}) = 0.0676\frac{1+2\,\mathrm{e}^{-\mathrm{j}\omega T}+\mathrm{e}^{-\mathrm{j}2\omega T}}{1-1.141\,\mathrm{e}^{-\mathrm{j}\omega T}+0.413\,\mathrm{e}^{-\mathrm{j}2\omega T}}$$

from which the magnitude squared is

$$|H(m)|^2 = 0.0092\frac{3+4\cos m\pi+\cos 2\,m\pi}{2.473-3.447\cos m\pi+0.826\cos 2\,m\pi} \tag{4.43}$$

where as before $m\pi = \omega T$. The function $10\log_{10}|H/H(0)|^2$ has been calculated for $0 \leq m \leq 2$ in steps of 0.05, using the computer program given in section 5 of the appendix. The result is shown in fig. 4.8, where the specification points are also inserted.

The bilinear z-transform, the impulse invariance, and other analogue-to-digital filter transformations can be obtained by other approaches. One particularly interesting approach

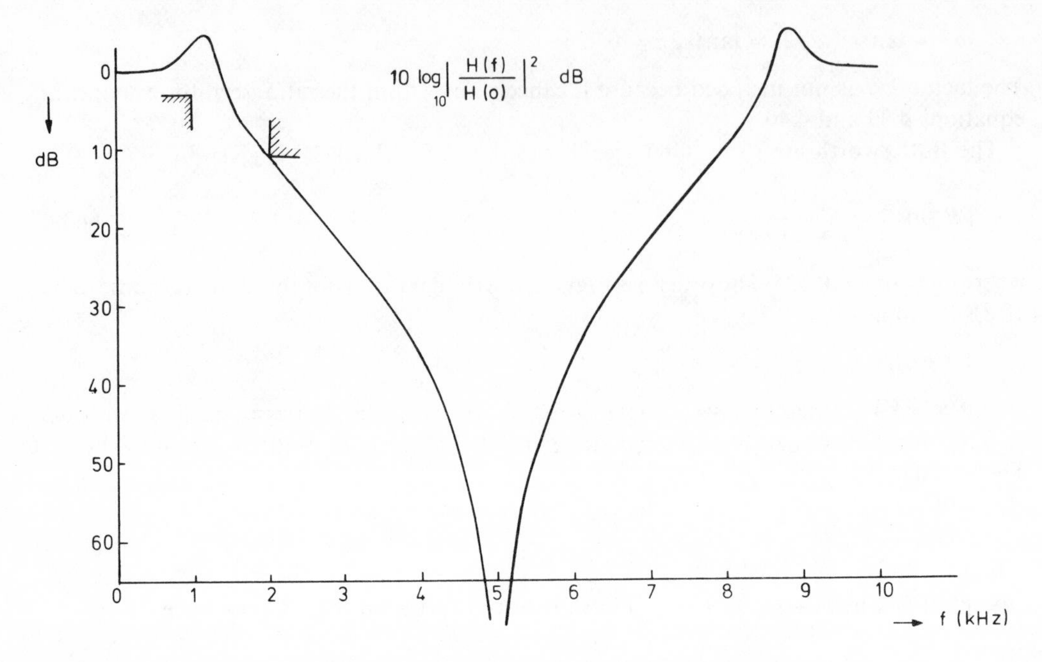

Fig. 4.8 Frequency response for digital second-order Butterworth filter

is based on the convolution integral expressed as an integro-difference equation, and is discussed by Haykin (18). Various transformations can then be obtained, depending on the specific type of approximation of the continuous-time input signal. There are also other interpretations (for example, as described by Oppenheim & Schaffer (10), section 5), and an interesting approach is also shown in problem 4.10.

4.3 Frequency transformations of lowpass filters

The design methods discussed in the previous sections, i.e. the impulse invariance and bilinear z-transform, perform transformation of lowpass analogue filter to digital lowpass filter. These transformations operate directly on the analogue filter variable s and produce the digital filter in variable z^{-1}. Other types of filters, such as highpass, bandpass and bandstop, can also be obtained by similar transformations of the lowpass analogue filter as discussed by Bogner & Constantinides (17), chapter 4.

Table 4.1 lists various transformations, from variable s to variable z^{-1}, for a given lowpass analogue filter of cutoff frequency Ω_c to a digital lowpass or highpass filter of cutoff frequency ω_c. They also enable transformation of a given lowpass to bandpass and bandstop filters, which are specified in terms of the upper (ω_2), lower (ω_1) and centre (ω_0) frequencies. A bandpass design example, based on Table 4.1, is given next.

Example 4.5
A digital signal processing system has a sampling rate of 1 kHz. Design a digital bandpass filter for the system with the following specifications:
(a) the range of the passband is from 100 to 400 Hz with ripple-free attenuation between 0 and 3 dB;

Table 4.1 Transformations from analogue filter variable s to digital filter variable

Required digital filter	*Replace* s *by:*	*Associated design parameters*
Lowpass	$\beta\dfrac{1-z^{-1}}{1+z^{-1}}$	$\beta=\Omega_c\cot\dfrac{\omega_c T}{2}$
Highpass	$\beta\dfrac{1+z^{-1}}{1-z^{-1}}$	$\beta=\Omega_c\tan\dfrac{\omega_c T}{2}$
		ω_c = desired cutoff frequency
Bandpass	$\beta\dfrac{z^{-2}-2\alpha z^{-1}+1}{1-z^{-2}}$	$\alpha=\cos\omega_0 T = \dfrac{\cos\dfrac{\omega_2+\omega_1}{2}T}{\cos\dfrac{\omega_2-\omega_1}{2}T}$
		$\beta=\Omega_c\cot\dfrac{\omega_2-\omega_1}{2}T$
Bandstop	$\beta\dfrac{1-z^{-2}}{z^{-2}-2\alpha z^{-1}+1}$	α = as above
		$\beta=\Omega_c\tan\dfrac{\omega_2-\omega_1}{2}T$
		ω_2, ω_1 = desired upper and lower cutoff frequencies; ω_0 = centre frequency

(b) attenuation must be at least 20 dB at 45 and 450 Hz, and must fall off monotonically beyond these frequencies.

For the bandpass filter, using the expressions given in Table 4.1, we have

$$\alpha=\cos\omega_0 T=\frac{\cos[(f_2+f_1)/f_s]\pi}{\cos[(f_2-f_1)/f_s]\pi}$$

$$\beta=\Omega_c\cot[(f_2-f_1)/f_s]\pi$$

Using values for f_1, f_2 and f_s as specified, we obtain

$$\alpha=\cos\omega_0 T=\frac{\cos\pi/2}{\cos 0.3\pi}=0 \tag{4.44}$$

$$\beta=\Omega_c\cot 0.3\pi=(0.72654253)\Omega_c \tag{4.45}$$

The transformation from s to z^{-1} plane, in this case, is

$$s=\beta\frac{1+z^{-2}}{1-z^{-2}} \tag{4.46}$$

Therefore, the analogue lowpass filter and digital bandpass filter frequencies are related by

$$\omega_a=-\beta\cot\omega_d T \tag{4.47}$$

Using this relationship we obtained the frequencies given in Table 4.2. Since filter response must be monotonic, we choose the Butterworth filter for which

$$|H(\omega)|^2 = \frac{1}{1+(\omega/\Omega_c)^{2n}} \tag{4.48}$$

where the cutoff $\Omega_c = 1$ (see for example Stanley (11)). The order of filter (n) is determined from the requirement that the above function is at least -20dB at 45 and 450 Hz.

Table 4.2 Digital bandpass and equivalent analogue lowpass frequencies

Digital bandpass	*Equivalent analogue lowpass*
$\omega_d = 2\pi 45$ rad s^{-1}	$\omega_{b1} = -2{\cdot}5501\ \Omega_c$
$= 2\pi 100$	$\omega_{a1} = -\Omega_c$
$= 2\pi 400$	$\omega_{a2} = -\Omega_c$
$= 2\pi 450$	$\omega_{b2} = 2{\cdot}2361\ \Omega_c$

Therefore we have

$$1+(2.23/1)^{2n} = 100$$

where $\omega_{b2} = 2.23$ rad s^{-1} has been taken as the more stringent requirement then $\omega_{b1} = -2.55$ rad s^{-1}.

The solution for n is then

$$n = 1/\log_{10} 2.23 = 2.87$$

hence we choose $n = 3$. The equivalent analogue lowpass (or prototype) filter is therefore third-order Butterworth with the transfer function

$$H(s) = \frac{1}{s^3+2s^2+2s+1} \tag{4.49}$$

The digital filter is then obtained using the substitution

$$s = 0.72654253\frac{1+z^{-2}}{1-z^{-2}} \tag{4.50}$$

in equation 4.49. The final expression for the digital bandpass filter transfer function is given as

$$H(z) = \frac{0.25691560\,(1-3z^{-2}+3z^{-4}-z^{-6})}{1-0.57724052\,z^{-2}+0.42178705\,z^{-4}-0.05629724\,z^{-6}} \tag{4.51}$$

where the numerical values have been calculated on a pocket calculator. The filter structure can be obtained, as before in example 4.2, by expressing $H = Y/X$.

To test the frequency response of the above filter, we have used the computer program given in section 5 of the appendix to calculate the function

$$10\log_{10}|H(m)/H(0.5)|^2 \text{ dB}$$

where $|H(m)|^2$ is equal to

$$0.51383120 \times \frac{10 - 15\cos 2m\pi + 6\cos 4m\pi - \cos 6m\pi}{1.51428032 - 1.68891710\cos 2m\pi + 0.90856820\cos 4m\pi - 0.11259448\cos 6m\pi} \tag{4.52}$$

for $m = 0, 0.05, 0.10, \ldots, 2$.

The result is plotted in fig. 4.9 where the specifications are also shown.

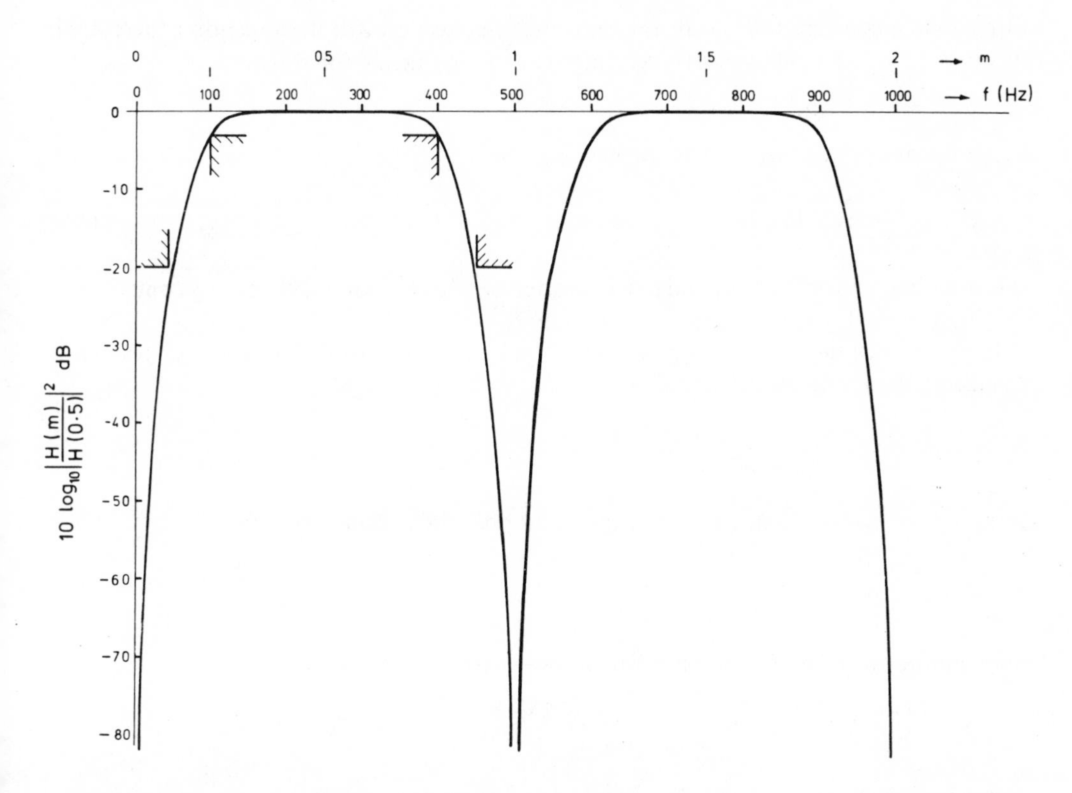

Fig. 4.9 Frequency response for digital bandpass filter

There are other transformations which transform a digital lowpass filter of cutoff frequency Ω_c into other digital (highpass, bandpass or bandstop) filters. These transformations operate on z^{-1} for the lowpass filter, which is replaced by an appropriate function $f(z^{-1})$ in order to obtain highpass, bandpass or bandstop filter (see, for example, Oppenheim & Schaffer(10), p. 230, and Bogner & Constantinides(17), chapter 5).

An interesting alternative approach to bandpass filter design is shown in example 4.6. This is based on the work by Smith *et al.* (19), which gives a more complete account of the design.

Example 4.6

The transfer function of a digital network given as

$$H(z) = \frac{a_0 z^2 + a_1 z}{z^2 + b_1 z + b_2} \tag{4.53}$$

is to have bandpass filter characteristics. Zeros of this equation are $z_1 = 0, z_2 = -a_1/a_0$, and poles are

$$p_{1,2} = -b_1/2 \pm \sqrt{(b_1^2/4 - b_2)}$$

For resonance to occur, $b_1^2/4$ must be less than b_2, in which case we obtain a complex-conjugate pair of poles

$$p_{1,2} = -b_1/2 \pm \mathrm{j}\sqrt{(b_2 - b_1^2/4)} = \sqrt{b_2}\exp[\mathrm{j}\cos^{-1}(b_1/2\sqrt{b_2})] \tag{4.54}$$

For stability, poles must be inside the unit circle (see section 2 of the appendix), therefore in this case $b_2 < 1$. Resonance is to take place at the frequency for which

$$\omega_0 T = \cos^{-1}(b_1/2\sqrt{b_2})$$

From the above, the resonant frequency is given by

$$f_0 = \frac{1}{2\pi T}\cos^{-1}(b_1/2\sqrt{b_2}) \tag{4.55}$$

It is seen that f_0 varies with b_1 and to a lesser degree with b_2, but it is effectively controlled by varying the clock frequency, i.e. by changing T.

In order to define the Q-factor, we need to transform the above roots from z-plane into s-plane by means of the relationship

$$s = \frac{1}{T}\ln z \tag{4.56}$$

Poles $p_{1,2}$ in z-plane then have corresponding poles in s-plane given by

$$s_{1,2} = \frac{1}{T}\ln p_{1,2}$$

Substituting for p_1 or p_2, we have for the real part of s_1 or s_2

$$\sigma_1 = \frac{1}{T}\ln\sqrt{b_2}$$

For a high-Q system, the 3 dB bandwidth ($= 2\Delta f$) is given by

$$2\Delta f = |\sigma_1|/\pi \tag{4.57}$$

see, for example, problem 4.11. Therefore in this case

$$2\Delta f = \frac{1}{\pi T}|\ln\sqrt{b_2}| \tag{4.58}$$

Since the circuit Q can be expressed also as $f_0/2\Delta f$, as in problem 4.11, we have

$$Q = \frac{\cos^{-1}(b_1/2\sqrt{b_2})}{2|\ln\sqrt{b_2}|} \tag{4.59}$$

Equations 4.58 and 4.59 enable us to calculate digital filter parameters b_1, b_2 to obtain the resonant circuit of desired bandwidth and Q-factor.

The hardware realization of this digital filter is shown in fig. 4.10(a). The delay units z^{-1} have been realized by means of bucket-brigades, shown in fig. 4.10(b), which enable circuit

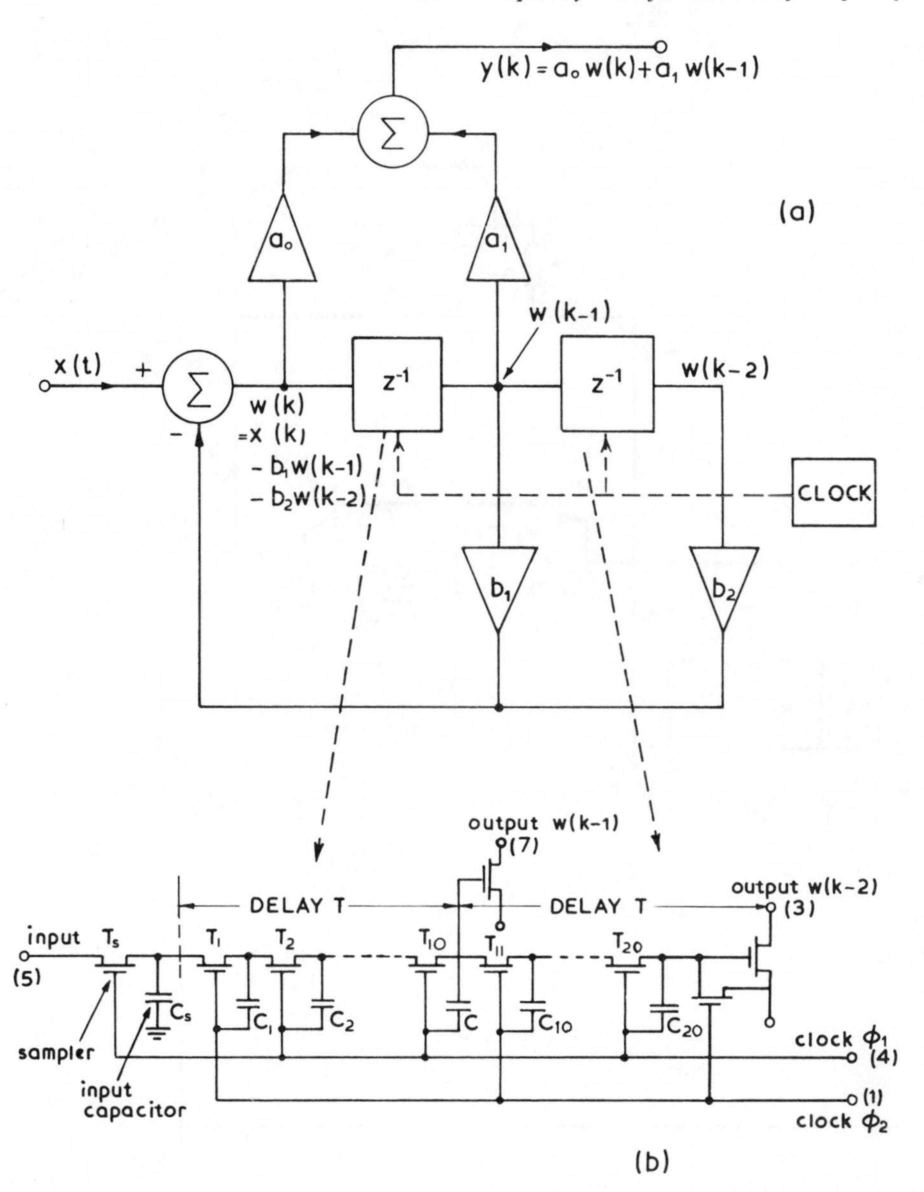

Fig. 4.10 (a) Hardware realization of discrete-time bandpass filter; (b) bucket-brigade delay units for use in (a)

operation with samples of analogue signals. The experimental circuit shown in fig. 4.11 uses a single integrated chip for the bucket delay line. The gains a_0, a_1, b_1, b_2 and the summing were implemented with operational amplifiers. Note that b_1 and b_2 can be adjusted by means of resistors R_1 and R_2 respectively. Results obtained with this circuit are well-documented by Smith *et al.* (19).

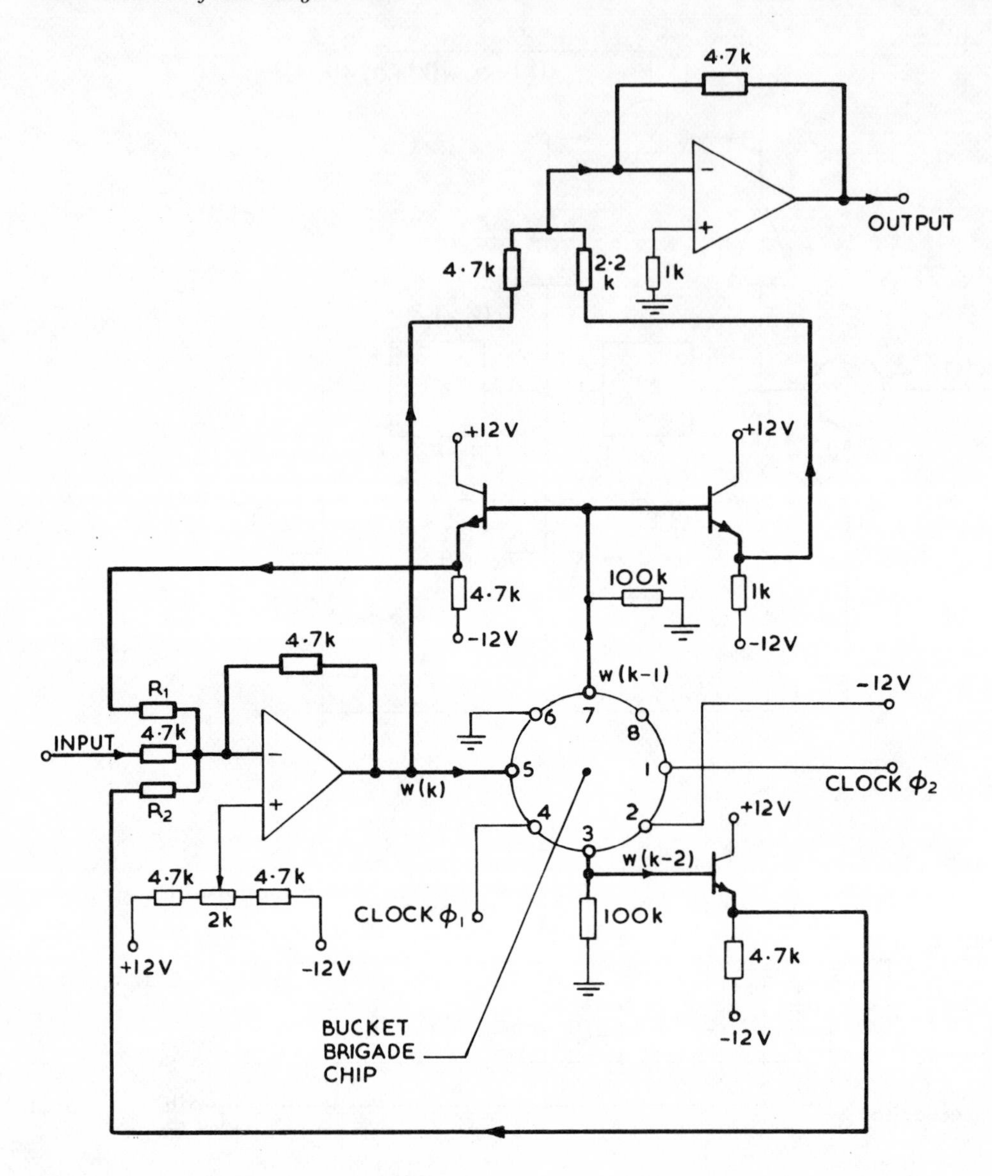

Fig. 4.11 Experimental circuit for fig. 4.10(a)

4.4 Accuracy considerations in digital filter design

In realizing a digital filter with either the physical hardware or a digital computer program, there will be errors introduced by the finite-word-length constraint. There are basically three types of error.

The first is caused by finite precision in the representation of the filter coefficients, which has an effect on both the desired transfer characteristic and filter stability. Nonrecursive filters have no feedback paths and hence no stability problem. The magnitude of their coefficient accuracy problem, and the transfer function, can be quickly estimated by simply looking at the relative magnitudes of the coefficients making up the weighting sequence. Since the recursive filter is a feedback structure, the problem of stability must be considered. An analysis of this problem by Kaiser(9) shows that the two design methods (impulse invariance and bilinear z-transform) have the same characteristic polynomial denominator of equation 2.4, in the limit as T is made small. This analysis also determines an absolute minimum bound on the number of decimal digits required to represent coefficients b_n in order to achieve filter stability, and examines the sensitivity of the denominator zeros to changes in the b_n. The main conclusions are as follows.

(i) If the filter being digitalized is of low order, having no lightly damped poles, and having poles of approximately the same modulus, and if the sampling rate is not much higher than five to ten times the average pole modulus, then most of the design procedures described in this chapter will work satisfactorily.
(ii) When the order of the filter is high, and when T is such that $p_n T \ll \pi$ (where p_n is the nth pole of $H(s)$) then care must be taken in the choice of design method.

The rule is to use the design method that involves (or at least permits) decomposition of the high-order filter into a group of low-order filters, as mentioned in section 2.4.

The second type of error results from the quantization of the input data to a specified number of bits, and the third type is due to the round-off in the multiplication and addition operations of the digital filter. To determine their effects on a digital filter, these sources of errors are treated as random noise sources, as described by Oppenheim & Schaffer(10) and Rabiner & Gold(15).

4.5 Problems

4.1 An analogue filter is specified in terms of a Butterworth second-order filter with the transfer function

$$H(s) = \frac{1}{s^2 + s/2 + 1}$$

where the cutoff is $\Omega_c = 1 \text{ rad s}^{-1}$.
(a) Derive a single digital filter transfer function using the impulse invariance method.
(b) From the result in (a) determine the filter structure in terms of the difference equation.

4.2 (a) For the analogue function given in problem 4.1, derive a single digital filter using the bilinear z-transform method.
(b) Determine the difference equation for this filter and compare it with the difference equation of problem 4.1 for $T \to 0$ (very frequent sampling).

4.3 A first order Butterworth filter normalized to $\Omega_c = 1 \text{ rad s}^{-1}$, is given by

$$H(s) = \frac{1}{s+1}$$

To normalize it to the cutoff frequency ω_c, we have to replace s by s/ω_c.
(a) Show how the corresponding impulse invariant digital filter is affected by the change of s to s/ω_c.
(b) How can the solution of problem 4.1 be changed to make it valid for the cutoff ω_c?

4.4 Design a lowpass digital filter with cutoff frequency ω_c, based on the analogue lowpass filter

$$H(s) = \frac{1}{s^2 + s/2 + 1}$$

which is normalized to $\Omega_c = 1 \text{ rad s}^{-1}$.
(a) Use Table 4.1 to find the solution in terms of β.
(b) Assuming that $\omega_c T/2$ is small so that $1/\beta \simeq \omega_c T/2$, derive the difference equation.
(c) Compare this solution with the one of problem 4.2. Note how ω_c enters the solution in this case.

4.5 (a) Design a single digital highpass filter of cutoff frequency ω_c based on the analogue lowpass filter as specified in problem 4.4. Compare the digital transfer function obtained for this filter with the one in problem 4.4. Can the highpass be obtained from the lowpass digital filter transfer function in a simple manner?
(b) Derive the highpass filter structure for the sampling frequency $f_s = 4f_c$. Note: It is advantageous to carry out the analysis in terms of β, and substitute for its value only in the last stages of the exercise.

4.6 Design a digital bandpass filter based on the lowpass analogue filter specified in problem 4.1. The passband range is to be from 100 to 700 Hz, and the system sampling frequency is 2.4 kHz.

4.7 Design a digital bandstop filter based on the lowpass analogue filter of problem 4.1. The stopband range is to be from 100 to 700 Hz, and the sampling frequency is 2.4 kHz.

4.8 The digital filters described by equations 4.16 and 4.37 are obtained respectively by application of the impulse invariant and bilinear z-transform methods to the first-order RC filter. Show that these two filters reduce to the same difference equation if sampling frequency is very large, in which case $\omega_c T \ll 1$.

4.9 (a) Apply the bipolar z-transform to the first-order analogue filter

$$H(s) = \frac{a}{s + a}$$

and show that the digital and analogue frequencies are related by equation 4.30.
(b) The impulse invariant transform of the above analogue filter is given by equation 4.11. Form the ratio $|H(\omega)/H(0)|$ for both the analogue and digital filter. Show that the analogue and digital frequencies are, in this case, related by

$$\omega_a = a\frac{\sin(\omega_d T/2)}{\sin(aT/2)}$$

(c) Show that for very frequent sampling, i.e. $T \to 0$, both results (a) and (b) reduce to $\omega_a = \omega_d$.

4.10 The Laplace and z-transform variables are related by $z = e^{sT}$ which can be written as

$$s = \frac{1}{T}\ln z$$

The ln z function can be approximated in the following three ways:

$$AP(1) = (z-1) - \tfrac{1}{2}(z-1)^2 + \tfrac{1}{3}(z-1)^3 - \ldots$$

$$AP(2) = \left(\frac{z-1}{z}\right) + \tfrac{1}{2}\left(\frac{z-1}{z}\right)^2 + \tfrac{1}{3}\left(\frac{z-1}{z}\right)^3 + \ldots$$

$$AP(3) = 2\left[\left(\frac{z-1}{z+1}\right) + \tfrac{1}{3}\left(\frac{z-1}{z+1}\right)^3 + \tfrac{1}{5}\left(\frac{z-1}{z+1}\right)^5 + \ldots\right]$$

see, for example, Abramowitz *et al.*(20), p. 68.
The exact value for $\ln(z = e^{j\omega T}) = j\,2\pi f/f_s$ where $f_s = 1/T$ is the sampling frequency. Assuming the sampling frequency $f_s = \lambda f$, where f is the highest frequency of interest in a given system, then we have $\ln z = j\,2\pi/\lambda$, and we define the error as

$$e_N = \left|\frac{\ln z - AP(N)}{\ln z}\right| \times 100\,\%$$

where $AP(N)$ represents the three approximations listed above. Taking only the first term in each $AP(N)$, calculate e_N, for $N = 1, 2, 3$, if $\lambda_1 = 2\pi \times 100$ and $\lambda_2 = 2\pi \times 10$. Note that the first term of $AP(3)$, i.e. $2[(z-1)/(z+1)]$, is the bilinear z-transform.

4.11 (a) Consider an analogue circuit consisting of R, C, L connected in parallel. Show that its input impedance, in the vicinity of resonance, is given by

$$Z(\omega \simeq \omega_0) = \frac{R}{1 + j2Q_0\Delta\omega/\omega_0}$$

where $Q_0 = R/\omega_0 L$ and $\omega_0 = 1/\sqrt{LC}$ are respectively the Q-factor and resonant frequency.
(b) From the above calculate the 3 dB bandwidth, and show that it can be expressed as

$$2\Delta f = |\sigma|/\pi$$

where $|\sigma| = 1/2RC$ is the real part of the poles of the input impedance calculated in the above part (a).
This problem is relevant to example 4.6.

5

Further concepts in discrete-time filtering

5.0 Introduction

In the first chapter we dealt with finite and infinite length discrete-time sequences, and introduced their z-transform representation. From this, the frequency spectrum has been obtained by the substitution $z = e^{j\omega T}$ as discussed in section 2.3.

In this chapter we return to the basic concepts, considering first a periodic discrete-time sequence and deriving its discrete Fourier series (DFS). Having established these relationships we show how the frequency spectrum of a finite (aperiodic) sequence is related to the periodic sequence frequency spectrum, and so obtain the discrete Fourier transform (DFT).

The other feature presented in this chapter is the inverse filter which shapes a finite input sequence into a unit pulse, or 'spike'. The least output error energy is then introduced as a criterion for derivation of the optimum finite length filter. This analysis is conducted in the time-domain which is convenient for solving some problems, for example in communications, radar, sonar and seismic explorations.

5.1 Derivation of discrete Fourier series (DFS)

This section uses some results derived in section 1.4. We established there two forms of frequency spectrum $F_s(s)$ for a sampled signal, expressed in terms of equations 1.17 and 1.20. Writing these two equations for $s = j\omega$, and equating them, we have

$$\sum_{n=-\infty}^{\infty} F(\omega + n\omega_s) = T \sum_{k=0}^{\infty} f(kT)e^{-jk\omega T} \tag{5.1}$$

The left-hand-side represents the periodic spectrum discussed in section 1.4. Denoting it here as $F_p(\omega)$ we obtain

$$F_p(\omega) = T \sum_{k=0}^{\infty} f(kT)e^{-jk\omega T} \tag{5.2}$$

For illustration we consider, in example 5.1, a simple discrete-time sequence.

Example 5.1

The sequence $f(kT) = e^{-akT}$, for $k > 0$, is shown in fig. 5.1(a) assuming $aT = 1$. It decays fast, therefore we can take for the upper limit $k = 5$, and calculate its frequency spectrum from equation 5.2 as

$$F_p(\omega) = T \sum_{k=0}^{5} e^{-kaT}e^{-jk\omega T}$$

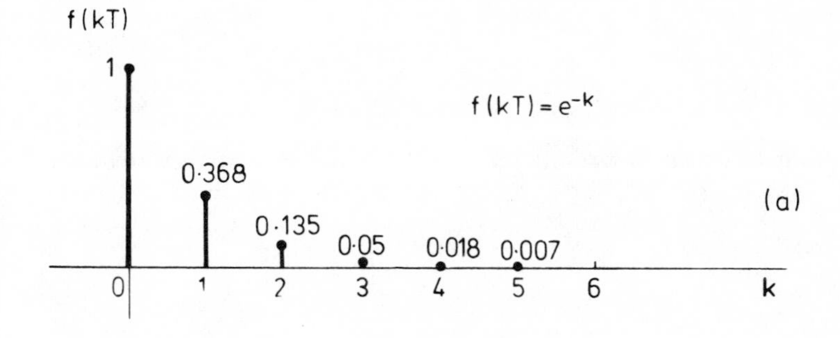

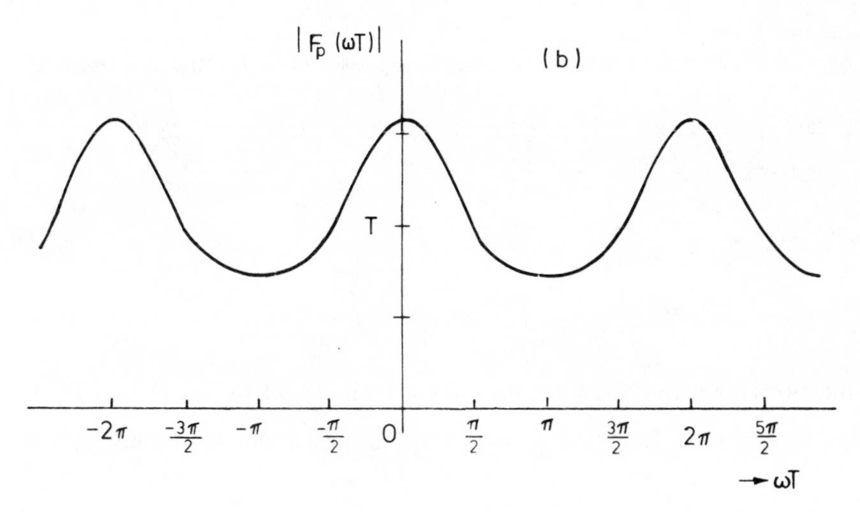

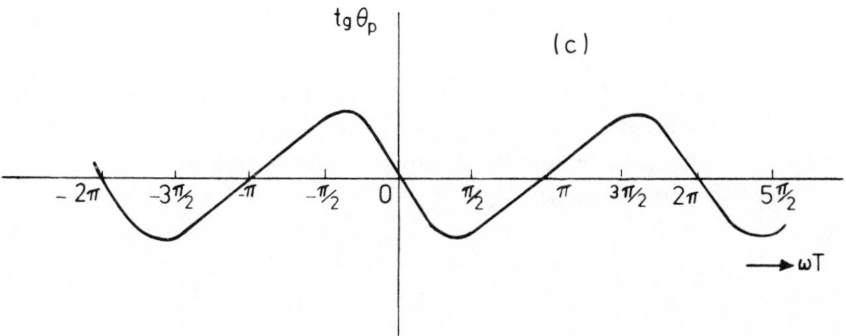

Fig. 5.1 (a) Decreasing time sequence; (b) magnitude and (c) phase of frequency spectrum of (a)

or $F_p(\omega) = T(1 + e^{-aT}e^{-j\omega T} + \ldots + e^{-5aT}e^{-j5\omega T})$

which is a finite geometric series. Summing this series gives

$$F_p(\omega) = T\frac{1 - e^{-6aT}e^{-j6\omega T}}{1 - e^{-aT}e^{-j\omega T}} \tag{5.3}$$

For $aT = 1$, $e^{-6aT} = e^{-6} = 0.0025$, and hence the numerator can be taken as unity, because also $|e^{-j6\omega T}| = 1$, so

$$F_p(\omega) = \frac{T}{1 - e^{-aT} e^{-j\omega T}} \tag{5.4}$$

It is interesting at this point to check this result for $T \to 0$, i.e. the case of very frequent sampling. We can then approximate

$$e^{-(a+j\omega)T} \simeq 1 - (a + j\omega)T$$

giving $F(\omega) = \dfrac{1}{a + j\omega}$

which is the familiar Fourier spectrum (aperiodic) for the exponential function $f(t) = e^{-at}$ where $t > 0$. This also shows that the spectrum relationships established earlier, equations 5.1 and 5.2, are correct.

The sampled-data spectrum given in equation 5.4 can be written as

$$F_p(\omega) = |F_p(\omega)| e^{j\theta_p}$$

where $$|F_p(\omega T)| = \frac{T}{(1 - 2e^{-aT} \cos \omega T + e^{-2aT})^{1/2}} \tag{5.5}$$

and $$\tan \theta_p = -\frac{e^{-aT} \sin \omega T}{1 - e^{-aT} \cos \omega T} \tag{5.6}$$

Both of these functions are periodic in frequency, as shown in fig. 5.1(b) and (c), for $aT = 1$.

Returning now to equation 5.2 and taking a finite length sequence with upper limit $(N - 1)$, we have

$$F_p(\omega) = T \sum_{k=0}^{N-1} f(kT) e^{-jk\omega T} \tag{5.7}$$

Assuming a periodic sequence $f_p(kT)$ with period $T_p (= NT)$ as shown in fig. 5.2, the frequency spectrum, which is already periodic because $f(t)$ is sampled, becomes discrete in frequency because $f(t)$ is periodic. We denote the discrete frequency spacing by $\Omega = 2\pi/T_p$, where T_p is the period of the time sequence.

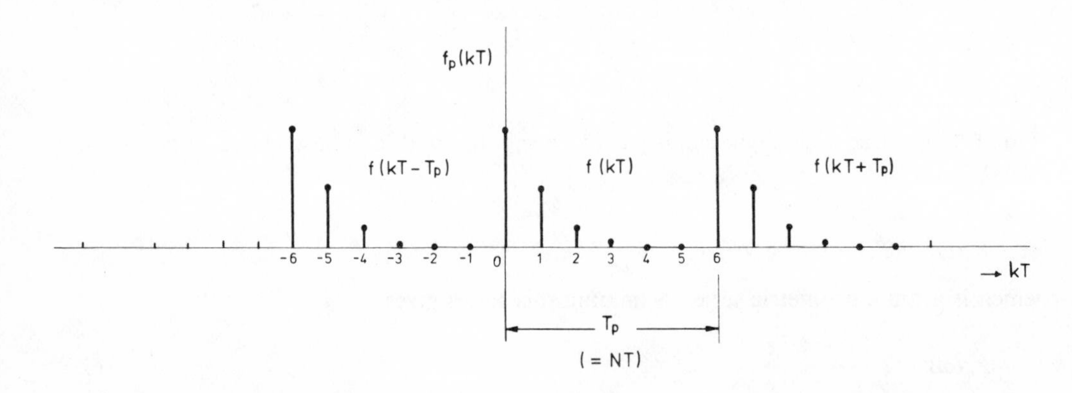

Fig. 5.2 Periodic sequence

The spectrum, equation 5.7, is now rewritten as

$$F_p(m\Omega) = T\sum_{k=0}^{N-1} f_p(kT)\,e^{-jkm\Omega T} \tag{5.8}$$

The next step is to derive an expression for $f_p(kT)$ in terms of the spectrum $F_p(m\Omega)$. For this purpose we multiply the above equation by $e^{jmn\Omega T}$ and sum over m as follows:

$$\sum_{m=0}^{N-1} F_p(m\Omega)\,e^{jmn\Omega T} = T\sum_{m=0}^{N-1}\sum_{k=0}^{N-1} f_p(kT)\,e^{-jkm\Omega T}\,e^{jmn\Omega T}$$

$$= T\sum_{k=0}^{N-1} f_p(kT)\sum_{m=0}^{N-1} e^{j\Omega T m(n-k)} \tag{5.9}$$

where we have taken the same number of samples (N) both in the time- and frequency-domain. Since $T_p = NT$, we have

$$\Omega = 2\pi/T_p = 2\pi/NT \tag{5.10}$$

and the second sum of equation 5.9 becomes

$$\sum_{m=0}^{N-1} e^{j(2\pi/N)m(n-k)} = \frac{1-e^{j2\pi(n-k)}}{1-e^{j(2\pi/N)(n-k)}} \tag{5.11}$$

which results in

$$\sum_{m=0}^{N-1} e^{j(2\pi/N)m(n-k)} = \begin{cases} N & n = k \\ 0 & n \neq k \end{cases} \tag{5.12}$$

where the first result follows from direct summation of the left-hand-side of equation 5.11, but the result for $n \neq k$ follows from the right-hand-side of equation 5.11 in which the numerator is then zero because $\exp[j2\pi(\text{integer})] = 1$. Returning to equation 5.9, we have

$$NT\sum_{k=0}^{N-1} f_p(kT) = \sum_{m=0}^{N-1} F_p(m\Omega)e^{j(2\pi/N)mk}$$

Therefore, the time sample at time kT is given by

$$NTf_p(kT) = \sum_{m=0}^{N-1} F_p(m\Omega)e^{j(2\pi/N)mk}$$

This result and equation 5.8, together with equation 5.10, form the following pair of equations:

$$F_p(m\Omega) = T\sum_{k=0}^{N-1} f_p(kT)e^{-j(2\pi/N)mk} \tag{5.13}$$

$$f_p(kT) = \frac{1}{NT}\sum_{m=0}^{N-1} F_p(m\Omega)\,e^{j(2\pi/N)mk} \tag{5.14}$$

to which we shall refer as the discrete Fourier series (DFS) for the periodic time sequence.

Although formulated for the periodic time-domain sequence, it will be seen in the next section that the same applies to finite (aperiodic) sequences.

5.2 Finite time sequences – discrete Fourier transform (DFT)

We have already met the expression

$$F_p(\omega) = \sum_{n=-\infty}^{\infty} F(\omega + n\omega_s) \qquad (5.15)$$

in section 1.4 and in equation 5.1 of the section 5.1. This equation shows the periodic nature of the frequency spectrum produced by sampling of a time-domain function. If the time function is periodic, then the frequency spectrum becomes discrete too. In this case, introducing $\omega = m\Omega$ and $\omega_s = N\Omega$ as discussed in the previous section, equation 5.15 can be rewritten as

$$F_p(m\Omega) = \sum_{n=-\infty}^{\infty} F(m\Omega + nN\Omega) \qquad (5.16)$$

Such a function is illustrated in fig. 5.3 (for $n = -1, 0, 1$), where we have assumed that the highest frequency in $F(m\Omega)$ is below $N\Omega/2(=\omega_s/2)$ so that the periodic sections do not overlap. Periodic spectra of this kind are of special interest since they are separable, as discussed in section 1.4.

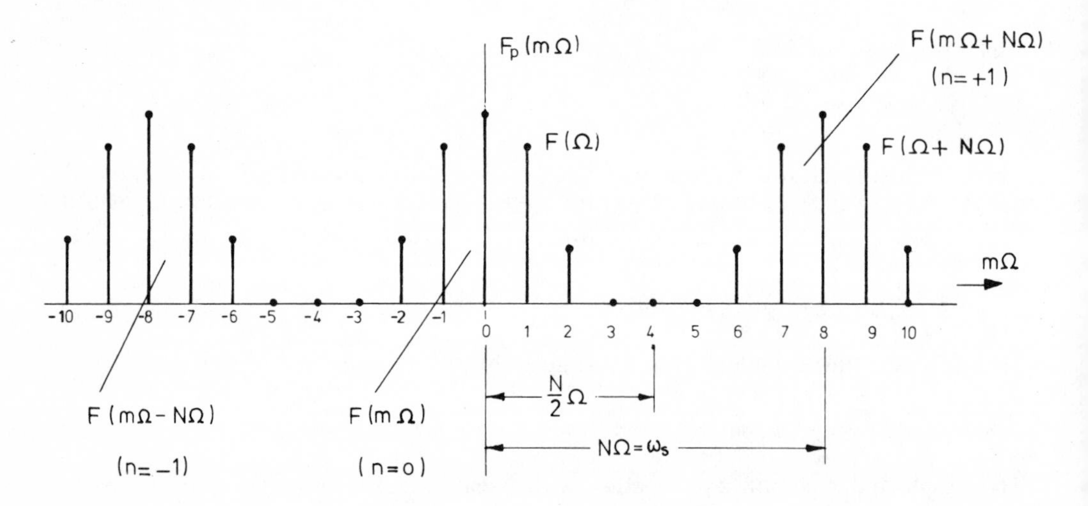

Fig. 5.3 Discrete frequency spectrum for a sampled function, periodic in time

Analogous to the above, a periodic and sampled time function can be expressed as

$$f_p(kT) = \sum_{n=-\infty}^{\infty} f(kT + nNT) \qquad (5.17)$$

where T is the discrete-time interval. Such a function, without overlap between successive periods, has been shown earlier in fig. 5.2. In such cases the periodic time function of equation 5.17 is assumed to be constructed from finite duration sequences repeated at intervals of NT. Conversely, in equations 5.13 and 5.14, $f_p(kT)$ represents values of the sequence within one period. We therefore deduce that the results for DFS, given by equations 5.13 and 5.14, are also applicable to the finite duration sequences, in which case they are written as

$$F(m\Omega) = T \sum_{k=0}^{N-1} f(kT)\,e^{-j(2\pi/N)mk} \qquad \text{where } 0 < m < N-1 \tag{5.18}$$

$$f(kT) = \frac{1}{NT} \sum_{m=0}^{N-1} F(m\Omega)\,e^{j(2\pi/N)mk} \qquad \text{where } 0 < k < N-1 \tag{5.19}$$

This pair of equations is known as the discrete Fourier transform (DFT). Note that although $f(kT)$ is a finite sequence, $F(m\Omega)$ is periodic, because $f(kT)$ is sampled while $F(m\Omega)$ is taken over one period.

We can also arrive at equation 5.18 by considering the z-transform of a finite duration sequence, i.e.

$$F(z) = \sum_{k=0}^{N-1} f(kT) z^{-k} \tag{5.20}$$

The sampled frequency response is then obtained by setting

$$z = e^{j\omega T}\Big|_{\omega = m\Omega} = e^{jm\Omega T} \tag{5.21}$$

and using $\Omega T = 2\pi/N$ from equation 5.10, we can rewrite equation 5.20 as

$$F(m\Omega) = \sum_{k=0}^{N-1} f(kT) e^{-j(2\pi/N)mk} \tag{5.22}$$

which is the same as equation 5.18. The factor T is missing in the above equation, because the z-transform given in equations 5.20 and 1.9 have been defined without T. This factor and similarly the spectral spacing Ω are often taken as unity in DFT and other similar expressions.

The computational aspects of DFT, and DFS, are illustrated in example 5.2.

Example 5.2

This example is an inversion of example 5.1, and it shows the main steps involved in calculating the time sequence from frequency sequence values. For six samples, i.e. $N = 6$, equation 5.14 gives

$$f_p(kT) = \frac{1}{6T} \sum_{m=0}^{5} F_p(m\Omega) e^{j(\pi/3)mk} \tag{5.23}$$

Using the expression for $F_p(m\Omega)$, derived as equation 5.4 in exercise 5.1, we have

$$F_p(m\Omega) = \frac{T}{1 - e^{-1} e^{-jm\pi/3}} \tag{5.24}$$

where we have taken $aT = 1$. It is convenient, for computational reasons, to represent the factor $e^{-jm\pi/3}$, for $m = 0$ to 5, on the unit circle diagram shown in fig. 5.4.

The time sample at $k = 0$ is given by

$$f_p(0) = \frac{1}{6T} \sum_{m=0}^{5} F_p(m\Omega)$$

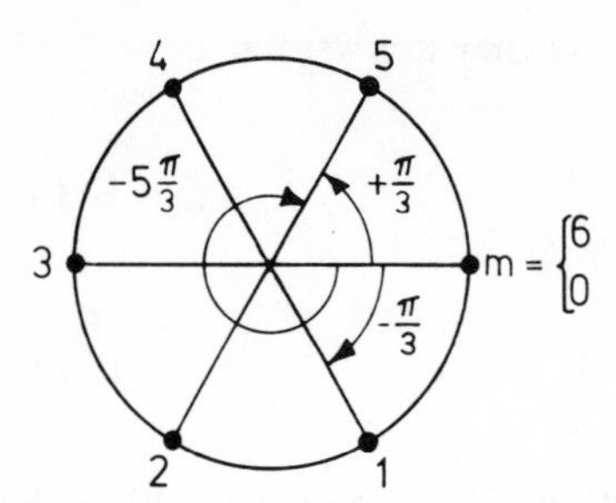

Fig. 5.4 Representation of $\exp(-jm\pi/3)$ on unit circle

Using $F_p(m\Omega)$ from equation 5.24, we have

$$f_p(0) = \frac{1}{6}\left(\frac{1}{1-e^{-1}} + \frac{1}{1-e^{-1}e^{-j\pi/3}} + \frac{1}{1-e^{-1}e^{-j2\pi/3}} + \frac{1}{1+e^{-1}} + \frac{1}{1-e^{-1}e^{-j4\pi/3}} + \frac{1}{1-e^{-1}e^{-j5\pi/3}}\right)$$

With reference to fig. 5.4, we see that terms one and five can be combined, and similarly terms two and four. The result of these operations is

$$f_p(0) = 2\left(\frac{1}{1-e^{-2}} + 2\frac{1-e^{-2}\cos(2\pi/3)}{1-2e^{-2}\cos(2\pi/3)+e^{-4}}\right) \tag{5.25}$$

which produces the value $f_p(0) = 1.09$. For the calculations, tables of exponential functions and a slide rule have been used. The actual value for $f_p(0)$ is unity, and the 'overshoot' is probably due to the discontinuity at the origin known as Gibbs' phenomenon. To check the method and accuracy of calculations, the sample $f_p(2)$ has also been calculated using the same procedure.

The time sample at $k = 2$ is given by

$$f_p(2) = \frac{1}{6T}\sum_{m=0}^{5} F_p(m\Omega)e^{j(2\pi/3)m}$$

The computational steps are similar to the ones used for $f_p(0)$, but now we have to take into account the factor $e^{j(2\pi/3)m}$, and diagrams similar to fig. 5.4 are very useful. The result is

$$f_p(2) = \frac{1}{3}\left(\frac{1}{1-e^{-2}} + \frac{e^{-1}-\cos(\pi/3)}{1-2e^{-1}\cos(\pi/3)+e^{-2}} - \frac{e^{-1}+\cos(\pi/3)}{1+2e^{-1}\cos(\pi/3)+e^{-2}}\right) \tag{5.26}$$

with the value $f_p(2) = 0.135$, which is in good agreement with the actual value of this time function as shown in fig. 5.1(a). We note that the time sequence in example 5.1 is taken as a finite duration sequence, but the time sequence components in example 5.2 are assumed to belong to a periodic time sequence as discussed earlier.

The above example is relatively simple but in general the numerical calculations are quite laborious, particularly if the number of samples N increases. The number of multiplications is N^2, i.e. N multiplications for each of the N frequency points. If N is a power of two, we need only $N \log_2 N$ multiplications. This efficient computational procedure is known as the fast Fourier transform (FFT), and it has made the DFT practical in many applications. For example if $N = 64 = 2^6$, we need only $N \log_2 N = 384$ multiplications, instead of $N^2 = 4096$. More detailed discussion of DFS, DFT, and FFT can be found in a number of books, for example, Oppenheim & Schaffer(10) and Rabiner & Gold(15).

5.3 Inverse filter

The process of inverse filtering or deconvolution is very useful in radar, sonar and seismic problems for removing undesirable components in a time series. These components may be caused by the medium through which the signal is passing, as illustrated in fig. 5.5, where g represents the impulse response of some propagation medium causing undesirable filtering or convolution described by $g * x$. The asterisk ($*$) denotes the convolution operation as given in equation 1.6.

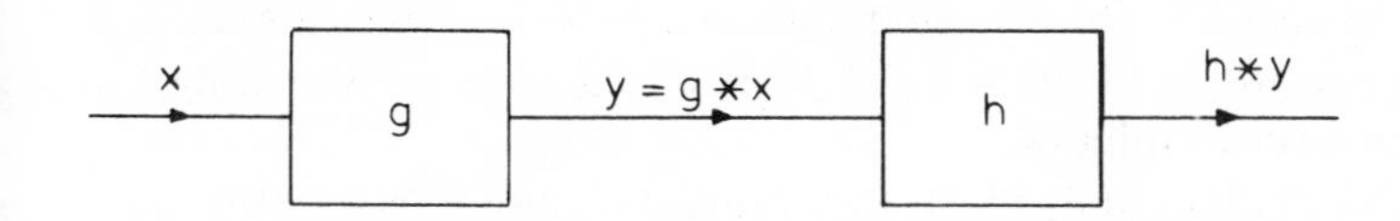

Fig. 5.5 Undesirable convolution by network g followed by deconvolution network h

The output of the following filter h is given by

$$h * y = h * g * x \tag{5.27}$$

We want to recover the signal x at the output of filter h. To achieve this we must find the filter h such that

$$h * g = \delta \tag{5.28}$$

where δ is the unit pulse sequence $[1, 0, 0, \ldots]$. Then the output of filter h is

$$h * y = \delta * x = x \tag{5.29}$$

i.e. the original signal x. To show that $\delta * x = x$ we can apply z-transform to both sides of this equation.

Applying z-transform to equation 5.28, we have

$$H(z)\,G(z) = 1 \tag{5.30}$$

as in, for example, problem 1.9. The required filter is therefore

$$H(z) = 1/G(z) \tag{5.31}$$

i.e. the required filter $H(z)$ is the inverse of the signal disturbing filter $G(z)$, sometimes denoted as $h = g^{-1}$. Assuming the unit pulse response of the disturbing filter is of arbitrary but finite length, we have

$$G(z) = g(0) + g(1)z^{-1} + \ldots + g(n)z^{-n} \tag{5.32}$$

The inverse filter of equation 5.31 is then given by

$$H(z) = \frac{1}{g(0) + g(1)z^{-1} + \ldots + g(n)z^{-n}}$$
$$= h(0) + h(1)z^{-1} + h(2)z^{-2} + \ldots \qquad (5.33)$$

where the h coefficients are obtained by long division. Therefore, the nonrecursive form of the inverse filter is of infinite length.

Example 5.3
Consider a special case of the disturbing filter described by $G(z) = 1 + az^{-1}$, i.e. $g(0) = 1$ and $g(1) = a$. The inverse filter in this case is given by

$$H(z) = \frac{1}{1 + az^{-1}} = \frac{z}{z + a}$$
$$= 1 - az^{-1} + a^2z^{-2} - a^3z^{-3} + \ldots \qquad (5.34)$$

which is obtained by the long division. The inverse (nonrecursive) filter coefficients are: $h(0) = 1$, $h(1) = -a$, $h(2) = a^2, \ldots, h(k) = (-a)^k$. If $a < 1$, these coefficients are decreasing, but theoretically this is an infinite series solution.

The above solution is checked graphically by means of the convolution relationship $g * h = \delta$ as shown in fig. 5.6. The input sequence $g(k)$ to filter h is regarded as a sum of pulse sequences, one, $g(0) = 1$ applied at $k = 0$, and another, $g(1) = a$ applied at $k = 1$. The responses to these two samples are shown in fig. 5.6, where the final output $\delta(k)$ is obtained by addition of the individual responses. Note that this system is linear and hence the principle of superposition holds.

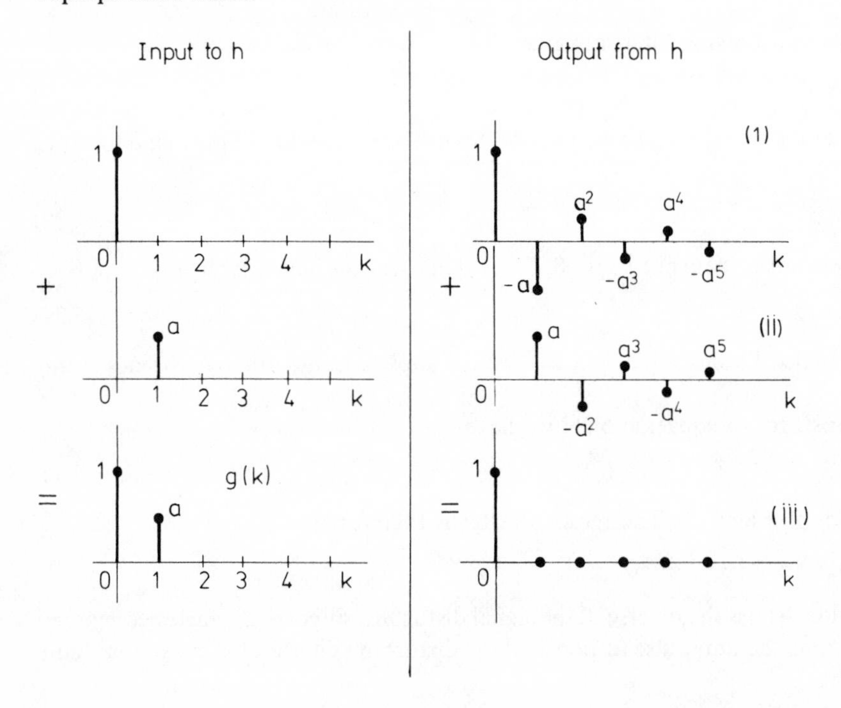

Fig. 5.6 (i) Response to 1; (ii) response to a; (iii) resultant response = $\delta(k)$

In practice we require finite length filters. Their output will be in error when compared to the desired output obtainable from an infinite length filter, and an analysis of finite length solutions and minimization of the error they cause are the subject of the following section.

5.4 Optimum finite inverse filter

The inverse filters introduced in the previous section are exact inverses, but they are infinitely long. In practice filters must have finite length, so they are only approximate inverses whose outputs will be in error. To develop the concept of error for finite or truncated filters we use the result of example 5.3, i.e.

$$\begin{aligned} g^{-1} = h &= [h(0), h(1), h(2), \ldots] \\ &= [1, -a, a^2, -a^3, \ldots] \end{aligned} \tag{5.35}$$

Consider first the one-length truncated inverse, i.e. $h = (\overset{\uparrow}{1})$. Convolving this approximate inverse with the input g, we obtain the actual output λ given by

$$\lambda = g * h = (1, a) * (1) = (1, a)$$

The desired output is the pulse ('spike') at time $k = 0$, represented by $\delta = (1, 0)$, so that the error between the desired and actual output is

$$e = \delta - \lambda = (1, 0) - (1, a) = (0, -a)$$

The sum of squares of the coefficients of the error sequence represents the error energy, in this case given by

$$0^2 + (-a)^2 = a^2$$

Consider now the two-length truncated inverse, i.e. $h = (1, -a)$. The actual output is given by

$$\lambda = g * h = (1, a) * (1, a) = (1, 0, -a^2)$$

which is easily obtained, for example, using the graphical method shown in fig. 5.6. The desired output is the pulse at time 0, i.e. $\delta = (1, 0, 0)$, and the error is

$$e = \delta - \lambda = (1, 0, 0) - (1, 0, -a^2) = (0, 0, a^2)$$

and its energy is a^4. This error energy is smaller than that for the one-length truncated inverse.

Results for one- to four-length truncated inverses are shown in fig. 5.7. We see that the error and its energy decrease as the length of the truncated (or approximate) inverse increases. However, it is possible to find an approximate inverse of a given length with smaller error energy than that of a truncated infinite inverse of the same length. To show this we derive one- and two-length inverses using the criterion of minimum error energy.

Consider first the one-length inverse $h = (h_0)$. In the truncated inverse case $h_0 = 1$, but now h_0 must be a parameter to optimize the filter in the sense of minimum error energy. The actual output is given by

$$\lambda = g * h = (1, a) * (h_0) = (h_0, ah_0)$$

and the desired output is $\delta = (1, 0)$. Therefore, the error is

$$e = \delta - \lambda = (1,0) - (h_0, ah_0) = (1 - h_0, -ah_0)$$

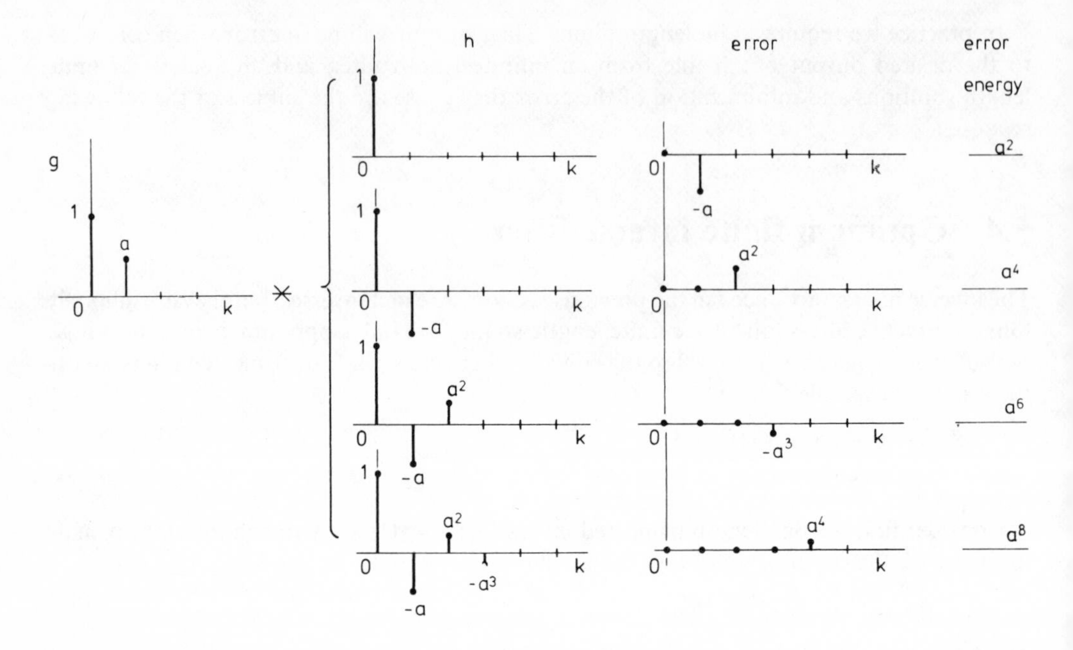

Fig. 5.7 Errors for one- to four-length truncated inverse filters

with energy

$$\begin{aligned} I &= (1-h_0)^2 + (-ah_0)^2 \\ &= 1 - 2h_0 + (1+a^2)h_0^2 \end{aligned}$$

We now minimize the above error energy with respect to h_0, i.e.

$$\frac{\partial I}{\partial h_0} = -2 + 2(1+a^2)h_0 = 0$$

which produces the solution

$$h_0 = \frac{1}{1+a^2} \tag{5.36}$$

for the one-length approximate inverse $h = (h_0)$.

We next consider the two-length inverse $h = (h_0, h_1)$, where again h_0 and h_1 are to be determined from the minimization of the error energy. The actual output in this case is

$$\lambda = g * h = (1, a) * (h_0, h_1) = (h_0, ah_0 + h_1, ah_1)$$

and the desired output is $\delta = (1, 0, 0)$. The error is

$$\begin{aligned} e = \delta - \lambda &= (1,0,0) - (h_0, ah_0 + h_1, ah_1) \\ &= (1 - h_0, -ah_0 - h_1, -ah_1) \end{aligned}$$

with the energy given by

$$\begin{aligned} I &= (1-h_0)^2 + (ah_0 + h_1)^2 + (ah_1)^2 \\ &= 1 - 2h_0 + (1+a^2)h_0^2 + 2ah_0h_1 + (1+a^2)h_1^2 \end{aligned}$$

To find the optimum values for h_0 and h_1 we minimize the above expression as follows:

$$\frac{\partial I}{\partial h_0} = -2 + 2(1+a^2)h_0 + 2ah_1 = 0$$
$$\frac{\partial I}{\partial h_1} = 2ah_0 + 2(1+a^2)h_1 = 0 \tag{5.37}$$

from which the optimum filter coefficients are

$$h_0 = \frac{1+a^2}{1+a^2+a^4} \tag{5.38}$$

and $$h_1 = \frac{-a}{1+a^2+a^4} \tag{5.39}$$

The corresponding minimum energy is

$$I_{\min} = \frac{a^4}{1+a^2+a^4} \tag{5.40}$$

which is smaller than the value a^4 obtained earlier by truncating the exact inverse to a two-length filter. Approximate inverses of greater length can be found by the method illustrated above for one- and two-length inverses. The optimum inverses so obtained are called the least error energy approximate inverses.

The least mean error energy (or least mean-square) criteria is used in many situations (see, for example, Robinson (8) and Schwartz & Shaw (21)). We shall be using this approach again in part 2 of the book. However, while we have dealt here with deterministic signals, we shall be dealing in part 2 with random signals, and the error energies (or squares) become the mean error energies (or squares).

5.5 Problems

5.1 Assume we have a finite length signal of duration 250 ms, sampled at 512 equally spaced points. Determine the following quantities of its discrete frequency spectrum:
(a) the increment in Hz between successive frequency components;
(b) the repetition period of the spectrum;
(c) the highest frequency permitted in the spectrum of signal to avoid interspectra (or aliasing) interference.

5.2 Calculate the frequency spectrum $F_p(m\Omega)$ for the periodic sequence shown in fig. 5.8, using equation 5.13, and write the z-transform, $F(z)$, for the first period ($k = 0$ to 3) of the sequence in fig. 5.8. Show that at points $\omega = m\Omega$, where Ω is given by equation 5.10, $F_p(m\Omega)$ and $F(z = e^{jm2\pi/N})$ are identical.

5.3 (a) Calculate the magnitude and phase of the frequency response for the periodic time sequence shown in fig. 5.9.
(b) Repeat the calculations of part (a) with the time origin in fig. 5.9 moved to $k = 2$. Compare the results obtained with the ones in part (a).

5.4 (a) Consider a discrete signal $f(k)$ whose first three samples are $f(0) = -\frac{1}{2}$, $f(1) = 1$, $f(2) = -\frac{1}{2}$. Compute its DFT using equation 5.18 and plot the magnitude of the frequency response $|F(m\Omega)|$ for $m = 0, 1, 2, 3$.

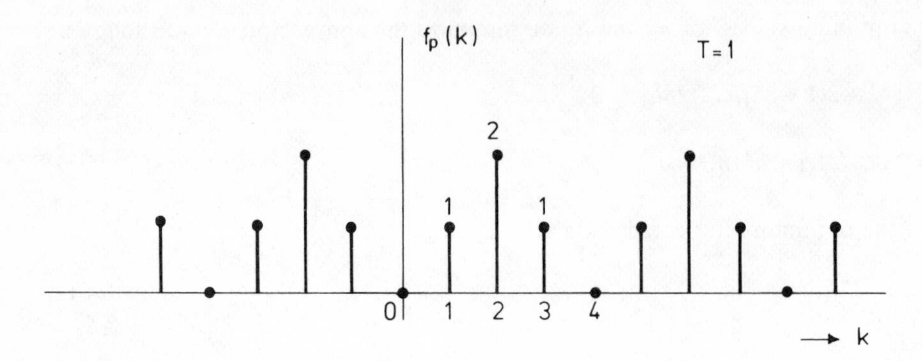

Fig. 5.8

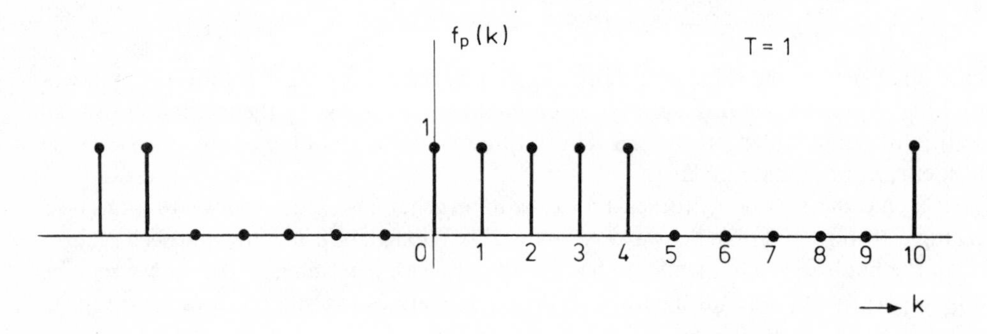

Fig. 5.9

(b) Add five zeros to the end of the data given in part (a), i.e. $f(3) = f(4) = \ldots = f(7) = 0$, and calculate the DFT of the new (augmented) sequence. Plot the magnitude $|F(m\Omega)|$ for $m = 0, 1, 2, \ldots, 8$, and show that the frequency resolution has increased. Note: This simple technique allows arbitrary resolution in computing the Fourier transform, as shown by Rabiner & Gold(15), pp. 54–5.

5.5 Given the sequence $g = (2, 1)$, (a) find the two-length and three-length approximate inverse and (b) compute the error squares for each case.

5.6 Compute the three-length approximate inverse for each of the sequences

$$g_1 = (2, 1, 1)$$
$$g_2 = (1, 1, 2)$$

and find the error energy for each case.

5.7 Show that the equations 5.37 can be written in the following generalized manner:

$$r(0)\,h_0 + r(1)\,h_1 = p(0)$$
$$r(1)\,h_0 + r(0)\,h_1 = p(1)$$

where the coefficients r and p are defined as follows (see, for example, Robinson(8),

p. 120): the coefficients r are the autocorrelations of the input sequence to the network h, defined in this case as

$$r(k) = \sum_{i=-\infty}^{\infty} g(i)\, g(i+k)$$

The coefficients p, are the cross-correlations of the input and desired output sequence of network h, defined in this case as

$$p(k) = \sum_{i=-\infty}^{\infty} g(i)\, d(i+k)$$

where $d(i)$ is desired output of network h: in our case $d = \delta = (1, 0)$.

Part 2 – Optimum (Wiener and Kalman) linear estimation

Introduction

This part of the book deals with extraction of signal from noisy measured data. The signal usually occupies a limited frequency range, while the noise is spread over a wide band of frequencies. In order to remove at least partly the noise from the signal we would use some kind of filtering, and in part 2, filters in terms of their discrete-time algorithms are used as the estimators of signals in noise. These filters, nonrecursive and first-order recursive, are the structures on which estimation theory is based. They are often referred to both as filters and as batch and sequential processors. The first area, nonrecursive or batch processing, is also known as classical estimation theory, while the second area, recursive or sequential processing, can be called modern estimation theory. The division of part 2 into three chapters does not strictly follow any of these areas, but reference is made to all of them.

To simplify the transition from part 1 to part 2, a brief review of digital filtering is given at the beginning of chapter 6, followed by a brief study of nonrecursive and first-order recursive filters as estimators reducing the noise in data. At this stage, we also introduce the concept of the mean-square error as a measure of the quality of noisy data processing by the chosen filter estimators. Here and in the following chapters, important equations such as those representing the estimator and corresponding mean-square error are underlined. In chapter 7, we develop the optimum nonrecursive and first-order recursive estimators based on minimization of the mean-square error. The results obtained are summarized in forms known as Wiener and Kalman filter algorithms.

All the material up to this point refers to the single signal which is termed the scalar (or one-dimensional) signal, but in chapter 8, we deal with the vector (or multidimensional) signals. Results obtained for the scalar signals are extended to vector signals using an equivalence between scalar and matrix operations.

Problems are not given at the end of each chapter of part 2, because the material is such that examples are either too simple or too complicated. However, throughout chapters 6 to 8, a number of examples are discussed in appropriate places. Also, there is a collection of suitable examples with solutions in chapter 9, which illustrate applications of the theory developed in earlier chapters.

6

Digital filtering of noisy data

6.0 Introduction

This chapter begins with a brief review of digital (i.e. discrete-time) filtering theory relevant to estimation theory as developed in the following chapters. It enables reading of part 2 without studying part 1, but more detailed information can be found in chapters 1 and 2. Further on in chapter 6, we consider the nonrecursive and first-order recursive filter structures as estimators of signals in noisy data. These particular filter structures are chosen since they lead into discrete-time Wiener and Kalman filter forms to be developed in chapter 7. The mean-square error is used as a criterion to assess the degree of noise suppression by the filter estimators.

6.1 Brief review of digital filtering

Digital filters are divided into two classes: nonrecursive and recursive types (see section 2.5). A nonrecursive filter structure is shown in fig. 6.1, where the input and output sampled-data signals are denoted by $f(k)$ and $g(k)$ respectively, but these are changed to y and $\hat{x}$ in the following sections when filters are considered as estimators. The output $g(k)$ is a finite

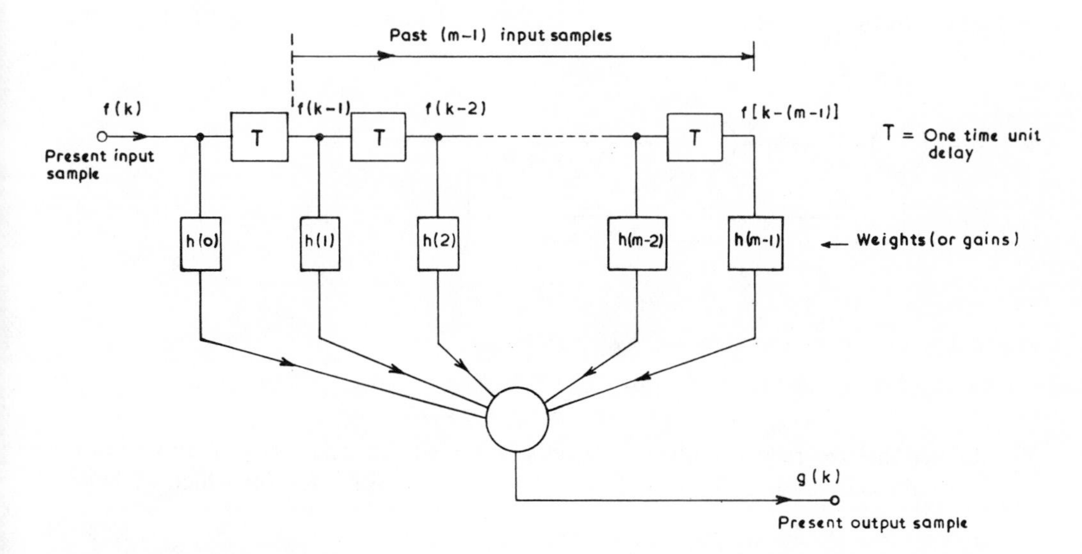

Fig. 6.1 Nonrecursive filter

weighted sum of the present input and a finite number of previous input samples. This is written as

$$g(k) = h(0)f(k) + h(1)f(k-1) + \ldots + h(m-1)f(k-m+1)$$

$$\text{or} \quad g(k) = \sum_{i=0}^{m-1} h(i)f(k-i) \tag{6.1}$$

which is the convolution summation in discrete-time (see equation 1.6). This represents a finite memory structure because it stores only a finite number m of input samples.

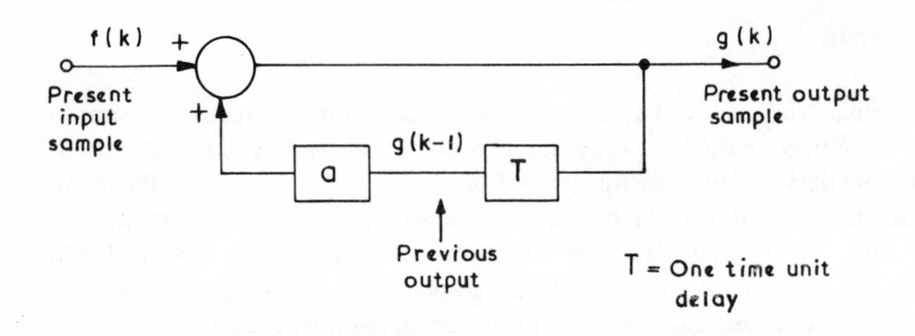

Fig. 6.2 Recursive (first-order) filter

The simplest first-order recursive filter structure is shown in fig. 6.2. This is a type of recursive structure used in the estimation theory covered in chapters 7 and 8. We notice immediately that the output, in this case, consists of the present input and a weighted previous output, where the weight is denoted by a. The output–input relationship can be written directly, by inspection of fig. 6.2, as

$$g(k) = f(k) + a\,g(k-1) \tag{6.2}$$

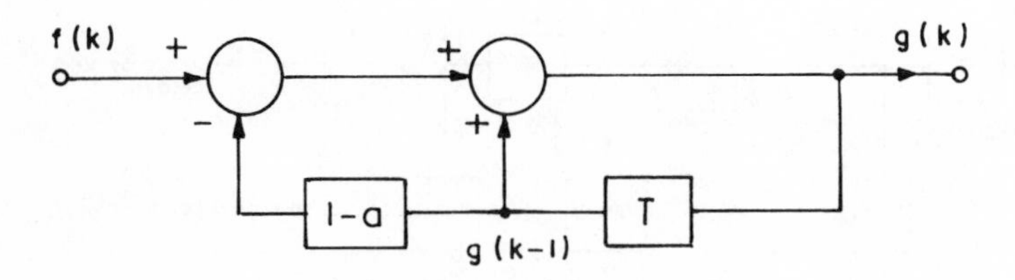

Fig. 6.3 An alternative form for fig. 6.2

We also see that the recursive filter is a feedback connection, and for stability we require $a < 1$. An alternative form of the recursive filter is shown in fig. 6.3, for which we have

$$g(k) = g(k-1) + [f(k) - (1-a)g(k-1)] \tag{6.3}$$

We shall come across this form in chapters 7 and 8.

Example 6.1
To show the basic difference between the nonrecursive and recursive filter, we use the unit-sample (or impulse) input defined as

$$f(k)=\begin{cases}0 & k<0\\ 1 & k=0\\ 0 & k>0\end{cases} \tag{6.4}$$

To simplify, we take $m = 3$, and obtain for the nonrecursive structure with $h(0) = 1, h(1) = \frac{1}{2}$, $h(2) = \frac{1}{4}$:

$$g(k) = f(k) + \tfrac{1}{2}f(k-1) + \tfrac{1}{4}f(k-2)$$

For the input specified above, we have

$$g(0) = 1,\ g(1) = \tfrac{1}{2},\ g(2) = \tfrac{1}{4},\ g(3) = g(4) = \ldots = 0 \tag{6.5}$$

which shows the finite impulse response (FIR) property.

Example 6.2
Applying the same input to the recursive structure given by equation 6.2, we obtain

$$g(0) = 1, g(1) = a,\ g(2) = a^2, \ldots, g(k) = a^k \tag{6.6}$$

where we have assumed $g(k) = 0$ for $k < 0$. We see that the recursive filter has an infinite impulse response (IIR), and it will decay only if $a < 1$.

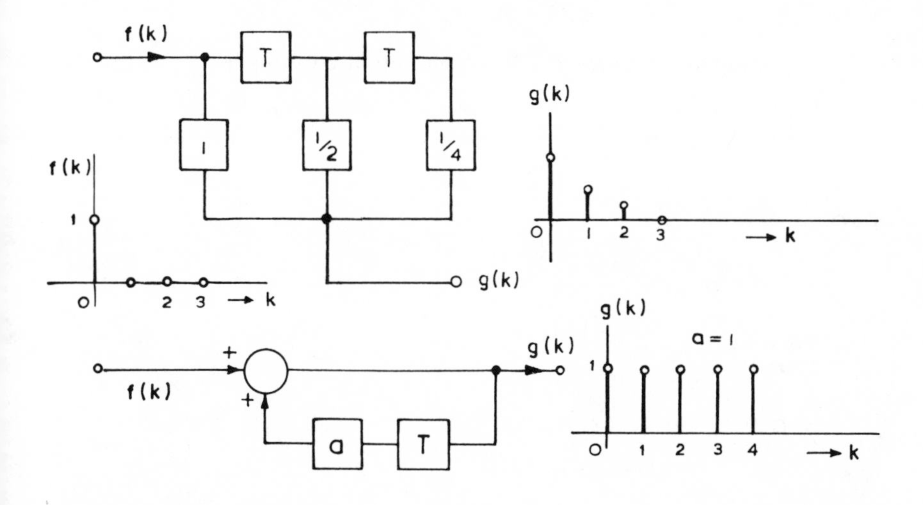

Fig. 6.4 Impulse response for nonrecursive and recursive filter examples

Results for examples 6.1 and 6.2 are shown in fig. 6.4.

To find the frequency response of these two structures we apply the discrete Fourier transform (see, for example, Schwartz & Shaw(21), p. 64) to both sides of equations 6.1 and 6.2. For illustration we consider the following two examples.

Example 6.3

To simplify analysis of the nonrecursive structure we take the case of a single storage element, i.e.

$$g(k) = \tfrac{1}{2}f(k) + \tfrac{1}{2}f(k-1) \tag{6.7}$$

where we have also assumed $h(0) = h(1) = \frac{1}{2}$. For this case we have

$$\sum_{k=-\infty}^{\infty} g(k)\,\mathrm{e}^{-\mathrm{j}k\omega T} = \tfrac{1}{2}\sum_{k=-\infty}^{\infty} f(k)\,\mathrm{e}^{-\mathrm{j}k\omega T} + \tfrac{1}{2}\sum_{k=-\infty}^{\infty} f(k-1)\,\mathrm{e}^{-\mathrm{j}k\omega T}$$

The last term can be written as

$$\tfrac{1}{2}\sum_{n=-\infty}^{\infty} f(n)\,\mathrm{e}^{-\mathrm{j}n\omega T}\mathrm{e}^{-\mathrm{j}\omega T} = \tfrac{1}{2}\mathrm{e}^{-\mathrm{j}\omega T}\sum_{n=-\infty}^{M} f(n)\,\mathrm{e}^{-\mathrm{j}n\omega T} = \tfrac{1}{2}\mathrm{e}^{-\mathrm{j}\omega T}F(\omega)$$

where we have substituted $n = k - 1$.

Therefore, the above equation can be written as

$$G(\omega) = \tfrac{1}{2}F(\omega)(1 + \mathrm{e}^{-\mathrm{j}\omega T})$$

Defining the transfer function as

$$H(\omega) = \frac{G(\omega)}{F(\omega)} = \tfrac{1}{2}(1 + \mathrm{e}^{-\mathrm{j}\omega T})$$

or $$H(\omega) = \mathrm{e}^{-\mathrm{j}\omega T/2}\cos \omega T/2 \tag{6.8}$$

we have

$$|H(\omega)| = \cos \omega T/2 \quad \text{for } 0 \le |\omega| \le \omega_s/2 \tag{6.9}$$

where $\omega_s = 2\pi/T$ is the sampling frequency. This result is plotted in fig. 6.5.

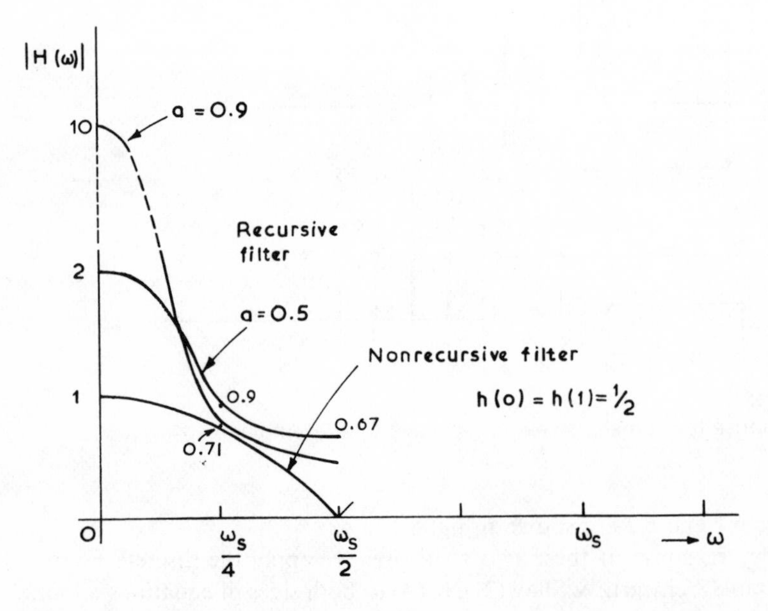

Fig. 6.5 Frequency response

Example 6.4
Applying the procedure used in example 6.3 to the recursive filter, we have

$$G(\omega) = F(\omega) + a\,e^{-j\omega T} G(\omega)$$

The transfer function is

$$H(\omega) = \frac{G(\omega)}{F(\omega)} = \frac{1}{1 - a\,e^{-j\omega T}} \tag{6.10}$$

from which the magnitude is

$$|H(\omega)| = \frac{1}{\sqrt{(1 + a^2 - 2a\cos\omega T)}} \quad \text{for } 0 \le \omega \le \omega_s/2 \tag{6.11}$$

This is plotted for $a = 0.5$ and $a = 0.9$, together with the result for the nonrecursive filter, in fig. 6.5. Note that the frequency responses, given by equations 6.9 and 6.11, are periodic functions of frequency and have been plotted up to $\omega_s/2$, (i.e. half of the sampling frequency).

It is seen from fig. 6.5 that the filters considered here are of the lowpass type. We shall not consider the frequency-domain any further, since, as stated in the introduction, estimation analysis is carried out in the time-domain. The above frequency-domain analysis has been done only to show that both structures behave as lowpass filters. The same analysis could have been done using z-transform, but this has been avoided since it would need more explanation than is required for the discrete Fourier transform described above.

6.2 Nonrecursive estimator

We adopt the notation x for a constant signal, and $x(k)$ for the time-varying signal. Measurement of this signal, denoted by $y(k)$, is linearly related to the signal x but has an additive noise component $v(k)$, introduced by random errors in measurements or any other causes. Therefore, we have

$$y(k) = x + v(k) \tag{6.12}$$

The signal considered here is a random variable with some mean value $E(x) = x_0$ and variance σ_x^2, where $E(x^2) = \sigma_x^2 + [E(x)]^2$ is denoted by S. The noise samples are assumed to be of zero-mean with identical variances σ_v^2, and also to be uncorrelated. This condition is not necessary but it is introduced to simplify analysis.

It is assumed that m data samples, as specified by equation 6.12, are to be processed using the nonrecursive filter structure of fig. 6.1, with all m weights equal to $1/m$. The input $f(k)$ is then $y(k)$, and the output $g(k)$ is taken as an estimate of parameter x, denoted by $\hat{x}$, as shown in fig. 6.6. It is important here to note that data $y(i)$, $i = 1, 2, \ldots, m$, are available as a *batch*. They are stored, multiplied by equal weights, and the result is summed to produce the output

$$\underline{\hat{x} = \frac{1}{m} \sum_{i=1}^{m} y(i)} \tag{6.13}$$

In general, the nonrecursive filter processor with different weights, is written as

$$\hat{x} = \sum_{i=1}^{m} h(i)\,y(i) \tag{6.14}$$

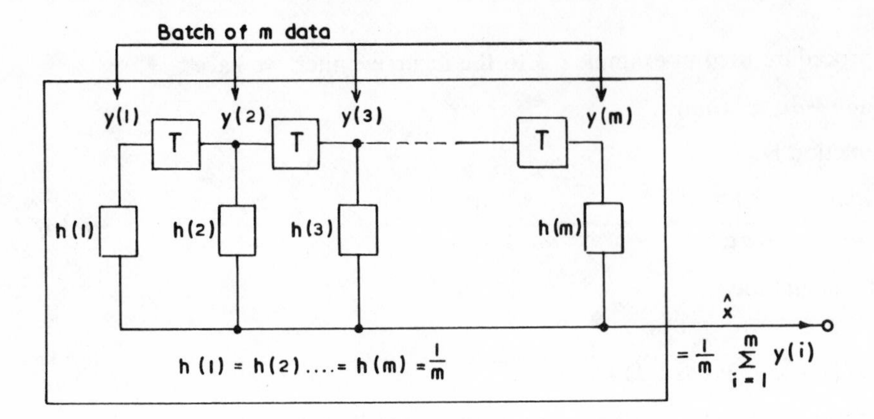

Fig. 6.6 Sample mean estimator

This is a modified convolution equation 6.1 for a finite length filter (m) and observation time $k > m$ (see, for example, Mendel (22), p. 15). The time index k does not appear since the time of data recording is of no importance in the processing of a batch. In fact, for the equal weights case of equation 6.13 it does not even matter in which order the data within a batch are taken. However, for the case of varying weights, as in equation 6.14, data have to be processed in an orderly time sequence.

The nonrecursive processor of equation 6.13 is the familiar *sample mean estimator* which we often use as a first approximation to estimate a quantity x from m data. Let us now assume that we want first to make an estimate of x from just one measured sample, say $y(i) = x + v(i)$, instead of m samples. This means we take $\hat{x} = y(i)$, and define the error between this estimate and actual value of x as $e = \hat{x} - x$. Then, the mean-square error is given by

$$p_e = E(e^2) = E(\hat{x} - x)^2 = E[x + v(i) - x]^2$$

$$\text{or} \quad p_e = E[v^2(i)] = \sigma_v^2 \tag{6.15}$$

We have introduced here the notation $E(\alpha)$ which represents the expected (mean or average) operation over the quantity (α). Now let us calculate the same for the estimate, over m data samples, as expressed by equation 6.13. In this case the mean-square error is given by

$$p_e = E(e^2) = E(\hat{x} - x)^2 = E\left\{\frac{1}{m}\sum_{i=1}^{m}[x + v(i)] - x\right\}^2$$

$$\text{or} \quad p_e = E\left\{\frac{1}{m}\left[\sum_{i=1}^{m} x + \sum_{i=1}^{m} v(i)\right] - x\right\}^2$$

$$= E\left[\frac{1}{m}\sum_{i=1}^{m} v(i)\right]^2$$

It is shown, in section 6 of the appendix, that the above becomes

$$\underline{p_e = \sigma_v^2/m} \tag{6.16}$$

which is an important relationship showing that as the number of samples m increases, the mean-square error p_e decreases. Therefore, the sample mean estimate is a good estimate of x in this sense.

The other interesting property of this estimator is obtained by taking the expectation of $\hat{x}$ in equation 6.13, which gives

$$E(\hat{x}) = E\left\{\frac{1}{m}\sum_{i=1}^{m} [x+v(i)]\right\} = E(x) = x_0 \tag{6.17}$$

since, as stated earlier in this section, $E(x) = x_0$, and $E[v(i)] = 0$.

The estimate of x, on average, is the same as the average of the estimate. Such an estimator is called an *unbiased estimator*, which, on average, produces the desired result.

6.3 Recursive estimator

We consider now the simple first-order recursive filter shown in fig. 6.7, where $y(k)$ and $g(k)$ are the input and output sequences respectively. The input signal $y(k)$ represents the measurements as expressed by equation 6.12, and the filter output is given by

$$g(k) = y(k) + ag(k-1) \tag{6.18}$$

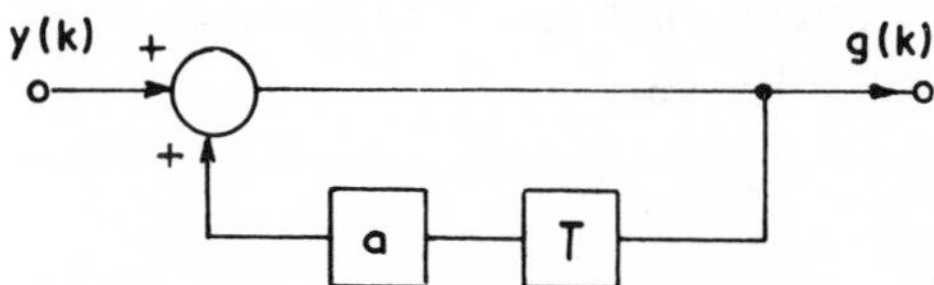

Fig. 6.7 Recursive filter as noisy data processor

i.e. this filter continually updates the output, adding a new data sample $y(k)$ to a fraction of the previous output $ag(k-1)$. To find the result of such a process we consider *sequentially* input samples $y(1), y(2), \ldots, y(m)$, assuming $g(k) = 0$ for $k < 1$. From equation 6.18 we have

$$g(0) = y(0) + ag(-1) = 0$$
$$g(1) = y(1) + ag(0) = y(1)$$
$$g(2) = y(2) + ag(1) = y(2) + ay(1)$$
$$g(3) = y(3) + ag(2) = y(3) + ay(2) + a^2y(1)$$

.

.

.

$$g(m) = y(m) + ag(m-1) = y(m) + ay(m-1) + a^2y(m-2) + \ldots + a^{m-2}y(2) + a^{m-1}y(1) \tag{6.19}$$

Substituting for $y(k) = x + v(k)$, and separating the signal and noise terms, we obtain

$$g(m) = (1 + a + a^2 + \ldots + a^{m-1})x + [v(m) + av(m-1) + \ldots + a^{m-1}v(1)]$$

or $$g(m) = \frac{1-a^m}{1-a}x + \sum_{i=1}^{m} a^{m-i}v(i) \tag{6.20}$$

where the first term is the sum of the geometric series associated with x. For large m, $|a|^m \ll 1$, and the signal part of $g(m)$ approaches $x/(1-a)$. This indicates that a good estimate of x is given by

$$\hat{x} = (1-a)g(m)$$

which leads to the important result

$$\hat{x} = (1 - a^m)x + (1 - a) \sum_{i=1}^{m} a^{m-i} v(i) \tag{6.21}$$

This means that the output $(1 - a)g(m)$ is taken as an estimate of signal x, after the mth input sample has been processed, as shown in fig. 6.8.

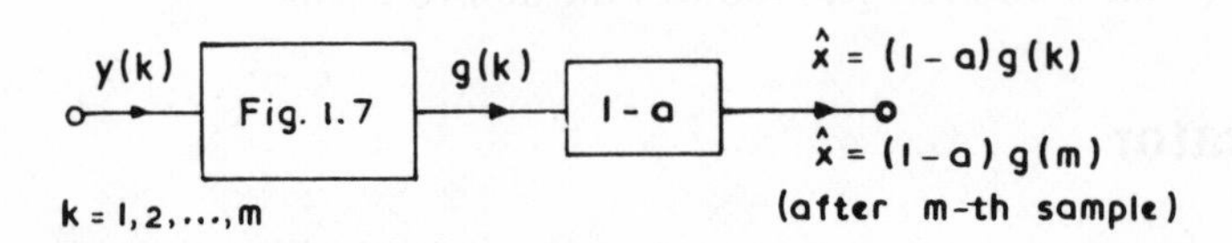

Fig. 6.8 Recursive filter as an estimator

To assess the accuracy of this *recursive estimator*, we calculate the mean-square error at the input and output. The input mean-square error, for one sample, is σ_v^2, as calculated in equation 6.15. The mean-square error at the output, after the mth sample has been processed, can be shown (section 7 of the appendix) to be given by

$$p_e = a^{2m} S + \frac{(1 - a)(1 - a^{2m})}{1 + a} \sigma_v^2 \tag{6.22}$$

which is another important result.

It is seen from equation 6.20 that $(1 - a)g(m)$, taken as an estimate of x, has two error components. The first one is due to a^m being nonzero, and the second is due to noise. We denote $\varepsilon = a^m$, and introduce

$$\lambda = \frac{1 + a}{1 - a} \frac{1}{1 - \varepsilon^2} \simeq \frac{1 + a}{1 - a}$$

since $\varepsilon^2 \ll 1$. Then equation 6.22 can be written as

$$p_e = \frac{\sigma_v^2}{\lambda}\left(1 + \frac{\lambda \varepsilon^2}{\gamma}\right) \tag{6.23}$$

where $\gamma = \sigma_v^2/S$. It is of interest to discuss this result briefly, taking three possible values of γ.

(i) If $\gamma \simeq 1$, i.e. $\sigma_v^2 = S$, then to reduce the output mean-square error we must make $\lambda \gg 1$. In this case we require $\lambda\varepsilon^2 \ll 1$ so that $p_e = \sigma_v^2/\lambda$. This is possible since $\varepsilon^2 = a^{2m}$ is a function of m, the number of samples. Therefore, for a fixed value of a satisfying $\lambda = (1 + a)/(1 - a)$, we have to choose sufficiently large m so that $\lambda\varepsilon^2 \ll 1$.

(ii) If $\gamma > 1$, i.e. $\sigma_v^2 > S$, it appears at first that we can satisfy $\lambda\varepsilon^2 \ll 1$ more easily because γ appears in the denominator. However, if $\sigma_v^2 > S$, then for a given S we have to increase λ in order to reduce σ_v^2/λ to an acceptable level. This in turn means that $\varepsilon^2 = a^{2m}$ has to be reduced, which can be done by increasing the number of data, i.e. increasing m.

(iii) If $\gamma < 1$, then $\sigma_v^2 < S$, so a smaller λ is required and a smaller m will be sufficient to make $\lambda\varepsilon^2/\gamma \ll 1$.

Example 6.5
Let $a = 0.5$ and $m = 4$, so that $\varepsilon = a^m = 1/16$. This means that $\hat{x}$ is on average within 6 % of x. The output mean-square error is

$$p_e = \frac{\sigma_v^2}{\lambda} = \frac{\sigma_v^2}{1.5/0.5} = \frac{\sigma_v^2}{3}$$

Example 6.6
If $a = 0.9$, we calculate the number of samples required for the same average 6 % error introduced in estimating x. From $\varepsilon = a^m$, we have for $\varepsilon = 1/16$

$$m = \frac{\log_{10}(1/16)}{\log_{10} a} = \frac{1.21}{0.045} = 27$$

The output mean-square error is $p_e = \sigma_v^2/\lambda = \sigma_v^2/19$. Here we have reduced p_e by increasing a, but to obtain a reasonable average estimate of x, we had to increase the number of samples from four, as calculated in example 6.5, to 27.

Frequency-domain analysis also helps in understanding the recursive filter processor discussed above. The noise is white, with a constant spectral density over all ω. The filter has a lowpass characteristic whose passband becomes narrower, letting less noise through, as the coefficient a takes values closer to unity (see, for example, fig. 6.5).

7
Optimum estimation of scalar signals

7.0 Introduction

We have seen in sections 6.2 and 6.3 that the mean-square error is a useful criterion showing how good an estimation process is. In this chapter, the mean-square error is taken as the fundamental criterion. The estimates that minimize the mean-square error are taken as the 'best' or optimum estimates, and are also referred to as the least mean-square (l.m.s.) estimates. The signal we consider in this chapter is a single signal which is named a scalar (or one-dimensional) signal. In sections 7.1 and 7.2 this is a constant signal parameter with a random distribution of its values. This is followed, in sections 7.3 and 7.4, by the case of a random time-varying single signal.

The optimum nonrecursive estimator derived in section 7.1 is the scalar Wiener filter whose coefficients are solutions of the Wiener–Hopf equation. Disadvantages of this batch type processor are noted in section 7.2, where a recursive type (or sequential) processor is developed from the optimum nonrecursive type. In section 7.3 we optimize the first-order recursive filter and group the results into the set of equations referred to as the scalar Kalman filter.

7.1 Optimum nonrecursive estimator (scalar Wiener filter)

In this section we deal with the nonrecursive filter whose output is to be the signal estimate, i.e.

$$\hat{x} = \sum_{i=1}^{m} h(i)y(i) \tag{7.1}$$

where $y(1), y(2), \ldots, y(m)$ are m data signals. This is a linear batch processor, briefly discussed in section 6.2. We assumed in that section that all coefficients have equal weights $(= 1/m)$. Now we want to choose these coefficients, $h(i), i = 1, 2, \ldots, m$, in such a way that the mean-square error

$$p_e = E(e^2) = E(x - \hat{x})^2$$

is minimized. Note that x is desired signal and $\hat{x}$ is its estimate, in this case given by equation 7.1. Substituting for $\hat{x}$, we have

$$p_e = E\left[x - \sum_{i=1}^{m} h(i)\,y(i)\right]^2 \tag{7.2}$$

The least (or minimum) mean-square error is obtained by differentiation of the above expression with respect to each of the m parameters. This can be written in the following concise form:

$$\frac{\partial p_e}{\partial h(j)} = -2E\left[x - \sum_{i=1}^{m} h(i)y(i)\right]y(j) = 0 \tag{7.3}$$

or $$\sum_{i=1}^{m} h(i)E[y(i)y(j)] = E[xy(j)] \quad \text{where } j = 1, 2, \ldots, m. \tag{7.4}$$

Before we proceed, it is worthwhile to note that from equation 7.3 we can also write

$$E[ey(j)] = 0 \quad \text{for } j = 1, 2, \ldots, m \tag{7.5}$$

where $e = x - \hat{x}$ is the error. This is called the *orthogonality principle* in discussions on estimation, and will be considered later in the book. It means that the product of the error $e = x - \hat{x}$ with each of the measured samples $y(j)$ is equal to zero in an expectation (or average) sense.

Returning to equation 7.4, we introduce

$$E[y(i)y(j)] = p_y(i, j) \tag{7.6}$$

which is, in fact, the data autocorrelation function usually denoted as $R_{yy}(i, j)$, but here we use p_y notation in agreement with the notation which is generally accepted in estimation theory, particularly Kalman filtering. It should also be noted that $p_y(i, j)$ refers to the non-stationary cases, while for the stationary cases it would be $p_y(j-i)$, see for example, Schwartz & Shaw (21), p. 278.

Similarly, we introduce

$$E[xy(j)] = p_{xy}(j) \tag{7.7}$$

which is the cross correlation between the random variables x and $y(j)$. Using equations 7.6 and 7.7, we rewrite equation 7.4 as

$$\sum_{i=1}^{m} h(i)p_y(i, j) = p_{xy}(j) \qquad \text{for } j = 1, 2, \ldots, m \tag{7.8}$$

which is an important result.

Expanding over $i = 1, 2, \ldots, m$, we have

$$p_y(1, j)h(1) + p_y(2, j)h(2) + \ldots + p_y(m, j)h(m) = p_{xy}(j)$$

and then expanding over $j = 1, 2, \ldots, m$, we obtain

$$\begin{aligned} p_y(1, 1)h(1) + p_y(2, 1)h(2) + \ldots + p_y(m, 1)h(m) &= p_{xy}(1) \\ p_y(1, 2)h(1) + p_y(2, 2)h(2) + \ldots + p_y(m, 2)h(m) &= p_{xy}(2) \\ &\vdots \\ p_y(1, m)h(1) + p_y(2, m)h(2) + \ldots + p_y(m, m)h(m) &= p_{xy}(m) \end{aligned} \tag{7.9}$$

where $p_y(i, j) = p_y(j, i)$ since function p_y is symmetrical. The *known* quantities are $p_y(i, j)$, i.e. the autocorrelation coefficients of the input data, and $p_{xy}(j)$, the cross-correlation coefficients between the desired output x and the input data y. The *unknown* quantities $h(i)$, $i = 1, 2, \ldots, m$, are the optimum filter coefficients. It is possible to take advantage of the special form of an autocorrelation matrix in order to reduce the computational work required (see, for example, Robinson(8), section 9.7).

The least mean-square error corresponding to the above optimum solution is obtained from

$$p_e = E(e^2) = E\{e[x - \sum_i h(i)y(i)]\} = E(ex)$$

on the basis of equation 7.5. Therefore, we have

$$p_e = E(x^2) - \sum_{i=1}^{m} h(i)E[xy(i)]$$

or $$p_e = E(x^2) - \sum_{i=1}^{m} h(i)p_{xy}(i) \tag{7.10}$$

where we have used equation 7.7 to establish this important result.

The complete solution to the estimation problem, in this case, is given by the set of equations 7.9, the estimate equation 7.1, and the corresponding least mean-square error given by equation 7.10.

Matrix forms of these three equations are

$$P_y\mathbf{h} = \mathbf{p}_{xy} \tag{7.9'}$$

where P_y is the $(m \times m)$ correlation matrix, $\mathbf{h}$ and $\mathbf{p}_{xy}$ are $(m \times 1)$ column vectors. The formal solution of equation 7.9′ is

$$\mathbf{h} = P_y^{-1}\mathbf{p}_{xy} \tag{7.11}$$

However, the estimate equation 7.1 can be written as

$$\hat{x} = \mathbf{h}^{\mathrm{T}}\mathbf{y} \tag{7.1'}$$

where $\mathbf{h}$ and $\mathbf{y}$ are $(m \times 1)$ column vectors, and $\mathbf{h}^{\mathrm{T}}$ is a row vector. Substituting equation 7.11 into 7.1′ we obtain for the estimate

$$\hat{x} = \mathbf{p}_{xy}^{\mathrm{T}}P_y^{-1}\mathbf{y} \tag{7.12}$$

and similarly for the least mean-square error

$$p_e = E(x^2) - \mathbf{p}_{xy}^{\mathrm{T}}P_y^{-1}\mathbf{p}_{xy} \tag{7.10'}$$

since the matrix P_y is symmetrical. Equations with form similar to 7.11 also arise in control theory where the least-square criterion is used for system identification, as discussed by Mendel(22), p. 67, and Eykhoff(23), chapter 2.

A filter of the type described above is often called a (scalar) *Wiener filter*, and equation 7.8 is known as the (scalar) *Wiener–Hopf equation.* The term scalar has been used here to distinguish this single signal case from the vector (multidimensional) case considered in section 8.5.

It should be noted that the relationship $y(k) = x + v(k)$, equation 6.12, has not been used in the above derivations. Therefore, the result is more general than it appears. It states that if the data samples $y(i)$, $i = 1, 2, \ldots, m$, 'somehow' contain the unknown random variable x, the signal, the best *linear*-filter operation carried out on the samples in order to estimate x is given by the Wiener filter (see, for example, Schwartz & Shaw (21), p. 279).

Example 7.1

Let measured data and signal be related linearly by $y(k) = x + v(k)$, where $v(k)$ is additive noise. We assume, as before, that the noise samples are of zero mean, with variance σ_v^2, uncorrelated with each other and with the signal x. This means that we assume white noise, so

$$E[v(j)v(k)] = \begin{cases} 0 & j \neq k \\ \sigma_v^2 & j = k \end{cases}$$

and $E[xv(j)] = 0$. Furthermore, to simplify, we assume $E(x) = 0$, and hence $E(x^2) = \sigma_x^2$.

To solve this problem, we calculate first

$$\begin{aligned} p_y(i,j) &= E[y(i)\,y(j)] = E\{[x+v(i)][x+v(j)]\} \\ &= \sigma_x^2 + \sigma_v^2\,\delta(i,j) \end{aligned} \tag{7.13}$$

where $\delta(i,j)$ is the Kronecker delta, i.e.

$$\delta(i,j) = \begin{cases} 0 & i \neq j \\ 1 & i = j \end{cases}$$

Next we have

$$p_{xy}(j) = E[x\,y(j)] = E(x^2) = \sigma_x^2 \tag{7.14}$$

Substituting equations 7.13 and 7.14 into the set of equations 7.9 we have

$$\left.\begin{aligned} &(\sigma_x^2+\sigma_v^2)h(1) + \sigma_x^2 h(2) + \ldots + \sigma_x^2 h(m) = \sigma_x^2 \\ &\sigma_x^2 h(1) + (\sigma_x^2+\sigma_v^2)h(2) + \ldots + \sigma_x^2 h(m) = \sigma_x^2 \\ &\quad\vdots \\ &\sigma_x^2 h(1) + \sigma_x^2 h(2) + \ldots + (\sigma_x^2+\sigma_v^2)h(m) = \sigma_x^2 \end{aligned}\right\}$$

or

$$\left.\begin{aligned} &\sigma_v^2 h(1) + \sigma_x^2 \sum_{i=1}^{m} h(i) = \sigma_x^2 \\ &\sigma_v^2 h(2) + \sigma_x^2 \sum_{i=1}^{m} h(i) = \sigma_x^2 \\ &\quad\vdots \\ &\sigma_v^2 h(m) + \sigma_x^2 \sum_{i=1}^{m} h(i) = \sigma_x^2 \end{aligned}\right\} \tag{7.15}$$

Summing both sides

$$(\sigma_v^2 + m\sigma_x^2) \sum_{i=1}^{m} h(i) = m\sigma_x^2$$

from which we have

$$\sum_{i=1}^{m} h(i) = \frac{m\sigma_x^2}{m\sigma_x^2 + \sigma_v^2}$$

Substituting this into each of equations 7.15, we obtain

$$h(1) = h(2) = \dots = h(m) = \frac{\sigma_x^2}{m\sigma_x^2 + \sigma_v^2} \tag{7.16}$$

Using the solution, equation 7.16, in equation 7.1, we have for the estimate

$$\hat{x} = \frac{1}{m+\gamma} \sum_{i=1}^{m} y(i) \tag{7.17}$$

where $\gamma = \sigma_v^2/\sigma_x^2$. The corresponding least mean-square error value, from equation 7.10 and 7.16, is given by

$$p_e = \frac{\sigma_v^2}{m+\gamma} \tag{7.18}$$

Note that for large signal to noise ratio we have $\gamma \ll m$, and the least mean-square estimator reduces to the sample mean estimate discussed in section 6.2.

Example 7.2
Assuming that two samples of a linearly increasing signal have been measured in the presence of independent additive noise, estimate the slope x of this straight line by means of the optimum linear processor

$$\hat{x} = \sum_{i=1}^{2} h(i)\, y(i)$$

Assume noise samples to be uncorrelated, with variance σ_v^2.
We can write the data equation as

$$y(k) = kx + v(k) \quad \text{for } k = 1, 2 \tag{7.19}$$

where kx is a straight line and $v(k)$ is additive noise. To solve the problem, we have to determine $p_y(i,j)$ and $p_{xy}(j)$ required in the set of equations 7.9.
From equation 7.6 we have

$$\begin{aligned} p_y(i,j) &= E[y(i)\,y(j)] = E\{[ix + v(i)][jx + v(j)]\} \\ &= i,j\,E(x^2) + E[v(i)\,v(j)] \\ &= i,jS + \sigma_v^2\,\delta(i,j) \qquad \text{for } i,j = 1, 2 \end{aligned} \tag{7.20}$$

where $\delta(i,j)$ is the Kronecker delta defined in example 7.1, and $E(x^2) = \sigma_x^2 + [E(x)]^2$ is denoted by S.
From equation 7.7, we have

$$\begin{aligned} p_{xy}(j) &= E[xy(j)] = E\{x[jx + v(j)]\} \\ &= j\,E(x^2) = j\,S \qquad \text{for } j = 1, 2 \end{aligned} \tag{7.21}$$

Applying 7.20 and 7.21 to equations 7.9, we obtain

$$\begin{aligned} (S + \sigma_v^2)\,h(1) + 2\,S\,h(2) &= S \\ 2\,S\,h(1) + (4\,S + \sigma_v^2)h(2) &= 2\,S \end{aligned}$$

Solutions of these equations are

$$h(1) = 1/(5+\gamma), \qquad h(2) = 2/(5+\gamma)$$

where again $\gamma = \sigma_v^2/S$. Therefore, the estimate of x from equation 7.1 is given by

$$\hat{x} = \frac{y(1)+2y(2)}{5+\gamma}$$

and the least mean-square error from equation 7.10 is

$$p_e = S\frac{\gamma}{5+\gamma}$$

7.2 Recursive estimator from the optimum nonrecursive estimator

The difficulties with the Wiener filter, derived in the previous section, are as follows:

(i) it requires previous knowledge (or stored estimates) of the autocorrelation matrix P_y (see equation 7.11);
(ii) the number of data samples m to be used in the processing must be specified beforehand;
(iii) if m is changed for any reason (for example, more data may become available), the calculations must all be repeated;
(iv) it requires the inversion of the $(m \times m)$ matrix P_y. If m is large this can take substantial computer time.

To allow updating of the estimate as more information becomes available, and to save on digital processing cost, another processing scheme has been developed known as the *recursive* (or sequential) processor. This will be the main theme of the following sections, but first we derive a recursive algorithm from the nonrecursive solution obtained in the previous section. This serves only as an introduction, but a rigorous approach follows in section 7.3.

The problem is specified as before: given successive samples $y(k) = x + v(k)$, provide a linear estimator

$$\hat{x} = \sum_{i=1}^{k} h(i)\,y(i)$$

such that the mean-square error $p_e = E(x-\hat{x})^2$, is as small as possible.

We already know the nonrecursive solution to this problem, as derived in example 7.1, and the results are restated here, for k samples, as

$$\hat{x} = \hat{x}(k) = \sum_{i=1}^{k} h(i)y(i) \qquad \text{where } h(i) = \frac{1}{k+\gamma} \tag{7.22}$$

with the corresponding mean-square error

$$p_e = p(k) = E[x-\hat{x}(k)]^2 = \frac{\sigma_v^2}{k+\gamma} \tag{7.23}$$

where $\gamma = \sigma_v^2/\sigma_x^2$. It should be noted here that notation $\hat{x}(k)$ refers to the kth estimate of the parameter x, i.e. the estimate *after a batch of* k *samples* have been processed. Similar comment applies to $p(k)$ in equation 7.23. In the following section we deal with the time-varying signals $x(k)$, and then $\hat{x}(k)$ will mean the estimate of the signal process $x(k)$ *at time* k.

For $(k+1)$ samples the estimate and the corresponding mean-square error become

$$\hat{x}(k+1) = \sum_{i=1}^{k+1} h(i)\,y(i), \qquad h(i) = \frac{1}{(k+1)+\gamma} \tag{7.24}$$

$$p(k+1) = \frac{\sigma_v^2}{(k+1)+\gamma} \tag{7.25}$$

Since k represents the number of samples used, which is variable, it is more correct to write the coefficients as $h(i, k)$.

From equations 7.22 and 7.23, we have

$$h(i, k) = p(k)/\sigma_v^2$$

and similarly from equations 7.24 and 7.25

$$h(i, k+1) = p(k+1)/\sigma_v^2$$

Forming the ratio

$$\frac{p(k+1)}{p(k)} = \frac{h(i, k+1)}{h(i, k)} = \frac{k+\gamma}{k+1+\gamma} = \frac{1}{1+1/(k+\gamma)}$$

we obtain

$$\frac{p(k+1)}{p(k)} = \frac{1}{1+p(k)/\sigma_v^2} \tag{7.26}$$

This is a difference equation which, for given $p(k)$, can be solved for $p(k+1)$, then $p(k+2)$, etc. Therefore we have derived a simple algorithm for finding the variation of mean-square error with sample size.

Now consider the signal estimate $\hat{x}(k+1)$ obtained by processing $(k+1)$ samples. For this special case (example 7.1), we can write from equation 7.24

$$\hat{x}(k+1) = \frac{1}{k+1+\gamma} \sum_{i=1}^{k} y(i) + \frac{1}{k+1+\gamma} y(k+1)$$

and applying equation 7.22, we have

$$\hat{x}(k+1) = \frac{k+\gamma}{k+1+\gamma} \hat{x}(k) + \frac{1}{k+1+\gamma} y(k+1)$$

Using equation 7.26, we obtain

$$\hat{x}(k+1) = \frac{p(k+1)}{p(k)} \hat{x}(k) + \frac{p(k+1)}{\sigma_v^2} y(k+1) \tag{7.27}$$

This is a recursive estimating equation which together with equation 7.26 forms the required recursive algorithm. The procedure is to find $p(k+1)$, from equation 7.26, in terms of $p(k)$. Then, from the stored previous value $\hat{x}(k)$ and the new data sample $y(k+1)$, we can calculate $\hat{x}(k+1)$. This procedure *continually* generates the best linear mean-square estimator of x, and at the same time it provides the corresponding mean-square error, $p(k+1)$. Note that in this case, $p(k) \rightarrow 0$ for k very large.

To start this recursive process we must calculate the first estimate $\hat{x}(1)$, based on a single observation, by nonrecursive methods.

Example 7.3
Assume $\gamma = 2$, and σ_v^2 is given. Develop a sequence of recursive estimates of x.

To solve this problem, we find the first estimate of x from the nonrecursive estimator given by equation 7.22, which for the first data sample (i.e. $k = 1$), gives $\hat{x}(1) = \frac{1}{3}y(1)$. The first estimate, in this case, is equal to one third of the first data sample. The corresponding mean-square error, from equation, 7.23, is $p(1) = \frac{1}{3}\sigma_v^2$.

Now the recursion begins by using equation 7.26 to give

$$p(2) = \frac{p(1)}{1 + p(1)/\sigma_v^2} = \frac{[\frac{1}{3}\sigma_v^2]}{1 + \frac{1}{3}} = \frac{\sigma_v^2}{4}$$

Then, from equation 7.27, we have

$$\hat{x}(2) = \tfrac{3}{4}\hat{x}(1) + \tfrac{1}{4}y(2)$$

or since $\hat{x}(1) = \frac{1}{3}y(1)$,

$$\hat{x}(2) = \tfrac{1}{4}y(1) + \tfrac{1}{4}y(2)$$

which agrees with the nonrecursive method.

Similarly, we have $p(3) = \frac{1}{5}\sigma_v^2$, and the estimate after the third data sample is

$$\hat{x}(3) = \frac{p(3)}{p(2)}\hat{x}(2) + \frac{p(3)}{\sigma_v^2}y(3)$$

or $\quad \hat{x}(3) = \frac{4}{5}x(2) + \frac{1}{5}y(3)$

and so on.

Comparing the recursive relationship of equation 7.27 with the recursive filter, equation 6.18 in section 6.3, we note that it is of the same form, but with time-varying coefficients. Denoting these coefficients by

$$a(k+1) = \frac{p(k+1)}{p(k)} \qquad \text{and } b(k+1) = \frac{p(k+1)}{\sigma_v^2} \tag{7.28}$$

equation 7.27 becomes

$$\hat{x}(k+1) = a(k+1)\hat{x}(k) + b(k+1)\,y(k+1) \tag{7.29}$$

It is easily shown, using equations 7.28 and 7.26, that these parameters are related in the following way:

$$a(k+1) = 1 - b(k+1) \tag{7.30}$$

Therefore, equation 7.29 can also be written as

$$\hat{x}(k+1) = \hat{x}(k) + b(k+1)[y(k+1) - \hat{x}(k)] \tag{7.31}$$

The block diagrams for equations 7.29 and 7.31 are shown in figs 7.1 and 7.2 respectively.

The result, equation 7.31, is particularly interesting since it is in the form found in the following sections. It shows that the $(k+1)$th estimate is the same as the previous kth estimate plus a *correction term* depending on the difference between the new sample value $y(k+1)$ and the previous estimate. This correction term is multiplied by time-varying gain factor $b(k+1)$ which decreases with k, see equations 7.28 and 7.25.The estimate stabilizes ultimately at some value depending on the data sample, and will be modified only if a new sample $y(k+1)$ differs *considerably* from the previous estimate.

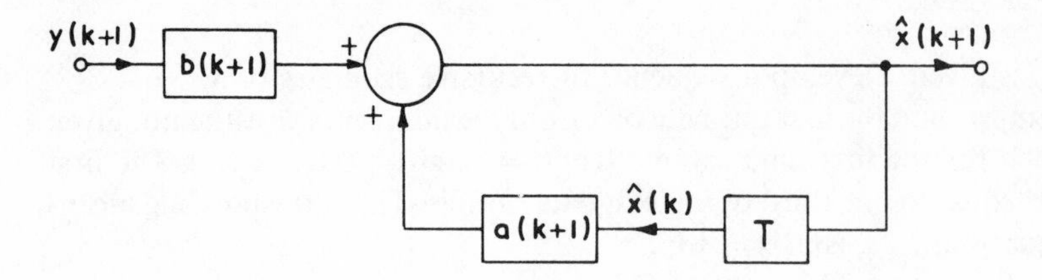

Fig. 7.1 Recursive filter (derived from nonrecursive estimator)

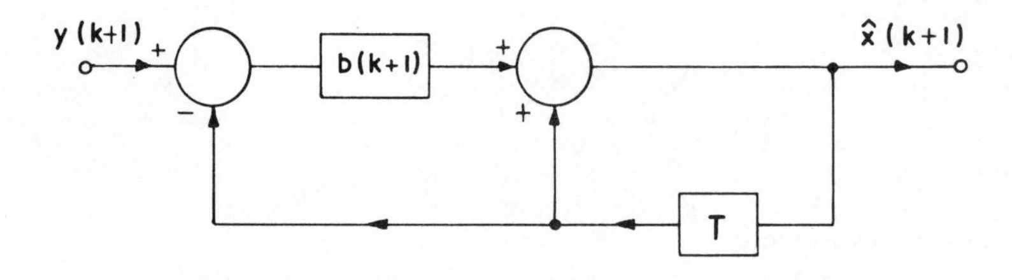

Fig. 7.2 An equivalent form of fig. 7.1

The recursive filter has been derived in this section from the nonrecursive filter solution for a particular case. It can be shown that the same result is obtained by starting with a recursive filter structure having two parameters $a(k+1)$ and $b(k+1)$ and using them to minimize the mean-square error of the estimate. This is the approach used in the following sections.

7.3 Optimum recursive estimator (scalar Kalman filter)

In this section we generalize the analysis in several ways.

(i) We deal with randomly time-varying signals or random processes. This is a common situation, for example, in radio transmission and aircraft tracking. Therefore, a model of such a signal will be discussed first.

(ii) The observation (data) equation is changed by a factor c multiplying the signal. We need it to enable the generalization of results to vector signals.

(iii) We derive the optimum estimate for a generalized first-order recursive filter. Results are arranged in so-called scalar Kalman filter form, suitable for a direct transformation into vector Kalman filter.

It is assumed that the process is stationary with time, i.e. the random signal generating mechanism does not vary with time.

7.3.1 Signal Model

We assume that the random signal to be estimated can be modelled as a *first-order recursive process* driven by zero-mean white noise. Therefore, the signal evolves in time according to the dynamical equation

$$x(k) = ax(k-1) + w(k-1) \tag{7.32}$$

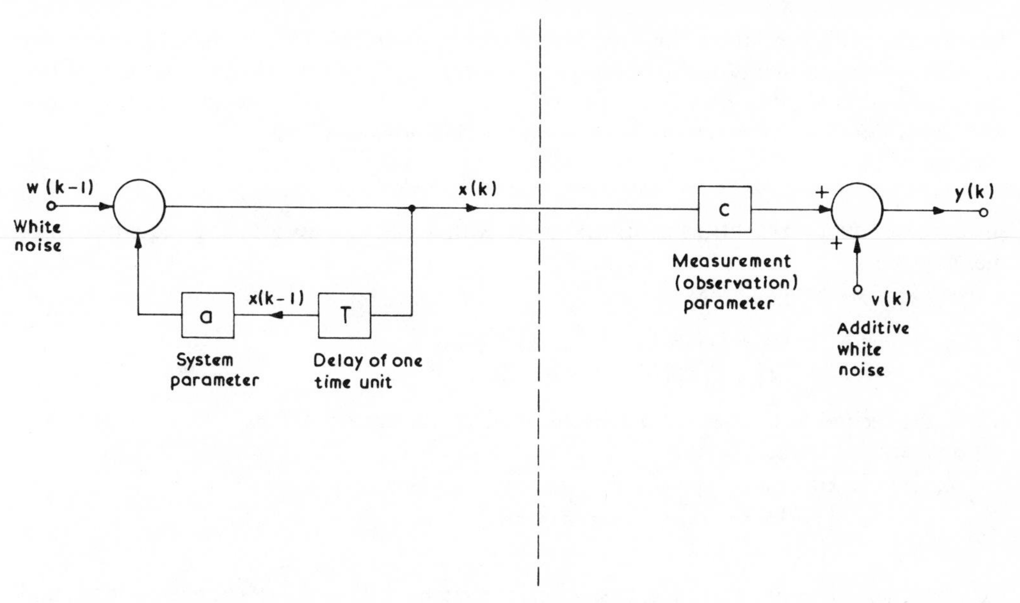

Fig. 7.3 Model of random signal process

Fig. 7.4 Measurement (observation) model

This model, shown in fig. 7.3, has $x(0) = 0$, if we assume the initial sample values to be zero, i.e. $x(k) = 0$ and $w(k) = 0$ for $k < 0$. The random drive is specified by

$$E[w(k)] = 0$$

$$E[w(k)w(j)] = \begin{cases} 0 & k \neq j \\ \sigma_w^2 & k = j \end{cases} \tag{7.33}$$

If $\sigma_w^2 = 0$, so that the white noise process disappears, and $a = 1$ with the initial condition $x(0) = x \neq 0$, we have $x(k) = x(k-1)$ which is a constant signal parameter x considered in previous sections.

A random process defined by equation 7.32 is said to be an *autoregressive process* of the first order. The statistical parameters of $x(k)$ are

$$E[x(k)] = 0$$

$$E[x^2(k)] = p_x(0) = \sigma_x^2 = \frac{\sigma_w^2}{1-a^2} \tag{7.34}$$

$$E[x(k)x(k+j)] = p_x(j) = a^{|j|}p_x(0) \tag{7.35}$$

where j represents the spacing between samples, and $p_x(j)$ is the autocorrelation function.

Example 7.4
Relationships 7.34 and 7.35 are derived as follows.

The autocorrelation function of two samples of $x(k)$ spaced j units apart is given by

$$p_x(j) = E[x(k)x(k+j)]$$

For $j = 0$, we have

$$p_x(0) = E[x^2(k)] = E[ax(k-1)+w(k-1)]^2$$
$$= a^2E[x^2(k-1)] + E[w^2(k-1)] + 2aE[x(k-1)w(k-1)]$$

so $\quad p_x(0) = a^2 p_x(0) + p_w(0)$

Introducing $p_x(0) = \sigma_x^2$ and $p_w(0) = \sigma_w^2$, we obtain equation 7.34. This result is valid assuming the filter to be in a steady-state condition and operating for a long time so that the output characteristics are stationary, i.e. $E[x^2(k)] = E[x^2(k-1)]$. To show that in the above derivation the third term is zero, we expand it using equation 7.32:

$$E[x(k-1)\,w(k-1)] = E\{[ax(k-2)+w(k-2)]\,w(k-1)\} = 0 \tag{7.36}$$

since we obtain a series of terms $E[w(k)w(j)]$ with $k \neq j$, and according to equation 7.33, these are zero.

For $j = 1$, we have

$$\begin{aligned} p_x(1) &= E[x(k)\,x(k+1)] = E\{x(k)[ax(k)+w(k)]\} \\ &= a\,E[x^2(k)] + E[x(k)\,w(k)] = a\,p_x(0) \end{aligned}$$

where the second term is zero for reasons as given for equation 7.36.

For $j = 2$, we have

$$\begin{aligned} p_x(2) &= E[x(k)\,x(k+2)] = E\{x(k)[a\,x(k+1)+w(k+1)]\} \\ &= a\,E[x(k)\,x(k+1)] + E[x(k)\,w(k+1)] \\ &= a\,p_x(1) = a^2\,p_x(0) \end{aligned}$$

etc., for other values of j. For j negative, we get the same result, so $p_x(j)$ is nonzero for all j and decreases with $|j|$ increasing. It will oscillate between positive and negative values when a is negative. The general result 7.35 follows from the above steps.

The parameter a plays the role of a *time-constant* for the process: the larger a is (approaching 1), the more sluggish the process is, requiring a larger time interval (in terms of units of sample spacing T) to change significantly from its current value. For more details see, for example, Schwartz & Shaw(21), pp. 333–4.

We assume again a *linear observation model* as shown in fig. 7.4, described by

$$y(k) = c\,x(k) + v(k) \tag{7.37}$$

where we have now a time-varying signal $x(k)$, and a factor c representing an observation or measurement parameter. It will be seen later that this factor is useful for the transformation of results to vector signals. As before, $v(k)$ represents an independent additive white noise with zero-mean and variance σ_v^2.

7.3.2 Optimum filter derivation

The *recursive estimator* is to be of the form

$$\hat{x}(k) = a(k)\,\hat{x}(k-1) + b(k)\,y(k) \tag{7.38}$$

where the first term represents the weighted previous estimate and the second term is weighted current data sample. Note that in the previous section $\hat{x}(k)$ was an estimate of the parameter x, *based on a batch of* k-*data*, but here $\hat{x}(k)$ is the estimate of the signal $x(k)$ at time k, based on the previous estimate and *only one data at time* k.

We want to determine the 'best' estimate using equation 7.38. The 'best' is again specified as the estimate which minimizes the mean-square error. In this case we have two parameters, $a(k)$ and $b(k)$, which are to be determined from minimization of the mean-square error

$$p(k) = E[e^2(k)] \tag{7.39}$$

where $e(k) = \hat{x}(k) - x(k)$ is the error. The same procedure was used in section 7.1, but the processor there was of nonrecursive type, and the number of coefficients $h(i)$ was equal to the number m of data available.

Substituting equation 7.38 for $\hat{x}(k)$, we have

$$p(k) = E[a(k)\hat{x}(k-1) + b(k)y(k) - x(k)]^2 \tag{7.40}$$

Differentiating with respect to $a(k)$ and $b(k)$ we obtain

$$\frac{\partial p(k)}{\partial a(k)} = 2E[a(k)\hat{x}(k-1) + b(k)y(k) - x(k)]\hat{x}(k-1) = 0 \tag{7.41}$$

and
$$\frac{\partial p(k)}{\partial b(k)} = 2E[a(k)\hat{x}(k-1) + b(k)y(k) - x(k)]y(k) = 0 \tag{7.42}$$

or alternatively

$$E[e(k)\hat{x}(k-1)] = 0 \tag{7.43}$$

and
$$E[e(k)y(k)] = 0 \tag{7.44}$$

which are *orthogonality equations* corresponding to equation 7.5 in section 7.1.

The first equation of the above set of equations is used in section 8 of the appendix to determine the relationship between $a(k)$ and $b(k)$

$$a(k) = a[1 - c\,b(k)] \tag{7.45}$$

Applying this to equation 7.38, we have

$$\underline{\hat{x}(k) = a\hat{x}(k-1) + b(k)[y(k) - ac\hat{x}(k-1)]} \tag{7.46}$$

The first term $a\hat{x}(k-1)$ represents the best estimate of $\hat{x}(k)$ without any additional information, and it is therefore a *prediction* based on past observations. The second term is a *correction* term depending on the difference between the new data sample and the observation estimate, $\hat{y}(k) = ac\hat{x}(k-1)$, with a variable gain factor $b(k)$; see fig. 7.5.

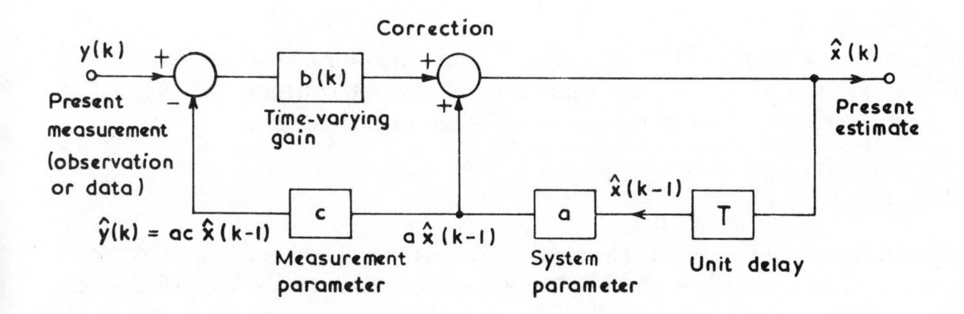

Fig. 7.5 Optimum recursive estimator (filter)

It is shown in section 9 of the appendix, that

$$\underline{b(k) = \frac{c[a^2 p(k-1) + \sigma_w^2]}{\sigma_v^2 + c^2\sigma_w^2 + c^2 a^2 p(k-1)}} \tag{7.47}$$

and also that the mean-square error is given by

$$p(k) = \frac{1}{c}\sigma_v^2\, b(k) \tag{7.48}$$

i.e. $p(k)$ and $b(k)$ are directly related. Note that equations 7.47 and 7.48, for $a = c = 1$ and $\sigma_w^2 = 0$, yield the result equation 7.26 derived in section 7.2.

Instead of starting the system with a nonrecursive estimate of $\hat{x}(1)$ based on first observation (as in example 7.3), we could attempt to make an estimate $\hat{x}(0)$ based on no observations. From this point of view, $\hat{x}(0)$ is the number which minimizes

$$p(0) = E[x(k) - \hat{x}(0)]^2$$

We have, then

$$\frac{\partial p(0)}{\partial \hat{x}(0)} = -2E[x(k) - \hat{x}(0)] = 0$$

or $$\hat{x}(0) = E[x(k)]$$

i.e. we find the best estimate $\hat{x}(0)$ is the mean value of $x(k)$.

Example 7.5

Using the above result, and the fact that we assumed $E[x(k)] = 0$, we can start the system as follows.

Since $\hat{x}(0) = 0$, and using the first data sample $y(1)$, we have from equation 7.46 $\hat{x}(1) = b(1)\,y(1)$.

To find $b(1)$ we use the orthogonality relation

$$E\{[x(1) - \hat{x}(1)]\,y(1)\} = 0$$

where $y(1) = x(1) + v(1)$, taking $c = 1$. Substituting for $\hat{x}(1) = b(1)\,y(1)$ and averaging we obtain

$$b(1) = \frac{\sigma_x^2}{\sigma_x^2 + \sigma_v^2}$$

Assuming further $\sigma_v^2 = \sigma_w^2$, and $a^2 = \frac{1}{2}$, we have from equation 7.34, $\sigma_x^2 = 2\sigma_v^2$, and $b(1) = \frac{2}{3} = 0.67$, in this case. We can use this value to calculate from equation 7.48, $p(1) = \frac{2}{3}\sigma_v^2$. Now we have all necessary data to use equation 7.47 to calculate $b(2)$ as

$$b(2) = \frac{a^2 p(1) + \sigma_w^2}{\sigma_v^2 + \sigma_w^2 + a^2 p(1)} = \frac{4}{7} = 0.57$$

for the numerical values given earlier. The mean-square error is $p(2) = 4\sigma_v^2/7$, from equation 7.48. Using $p(2)$ in equation 7.47 we have $b(3) = 9/16 = 0.562$, and then $p(3) = 0.562\,\sigma_v^2$, etc. As k increases, $p(k)$ reaches a steady-state value, i.e. it reaches the region where $p(k) = p(k-1) = p$. To find this value we substitute equation 7.48 into equation 7.47, and for this case obtain

$$p^2 + 3\sigma_v^2 p - 2\sigma_v^4 = 0$$

with the solution $p = 0.56\,\sigma_v^2$. We see that $p(3)$, calculated in the above, is already very close to the limiting (or steady-state) value of 0.56.

Equations 7.46 to 7.48 constitute a complete computational algorithm, and can be used in this way. However, for the purpose of extending these results to vector signals, in chapter 8, we rearrange these equations and write them below as equations 7.49 to 7.52.

Recursive filter estimator:

$$\hat{x}(k) = a\,\hat{x}(k-1) + b(k)[y(k) - a\,c\,\hat{x}(k-1)] \tag{7.49}$$

Filter gain:

$$b(k) = c\,p_1(k)[c^2 p_1(k) + \sigma_v^2]^{-1} \tag{7.50}$$

$$\text{where} \quad p_1(k) = a^2 p(k-1) + \sigma_w^2 \tag{7.51}$$

Mean-square error:

$$p(k) = p_1(k) - c\,b(k)\,p_1(k) \tag{7.52}$$

The estimating equation 7.49 remains the same as the original equation 7.46, but equations 7.47 and 7.48 are now written as three equations, 7.50 to 7.52, because we have introduced a new quantity, $p_1(k)$. This quantity has an important role, as will be seen in later sections. In fact, we should write $p(k|k-1)$ in place of $p_1(k)$ and $p(k|k)$ in place of $p(k)$; this is briefly discussed in section 8.3.

The above set of equations constitute the *scalar* (*one-dimensional*) *Kalman filter* for the signal model given by equation 7.32

$$x(k) = a\,x(k-1) + w(k-1)$$

and the measurement model given by equation 7.37

$$y(k) = c\,x(k) + v(k)$$

This type of recursive estimation was developed around 1960, most notably by R. E. Kalman (24). This is the reason for the processors devised at that time, as well as the wide variety of generalizations and extensions, to be referred to as *Kalman filters*. However, there were also other workers in this field who claimed priority, as discussed by Sorenson(25).

7.4 Optimum recursive predictor (scalar Kalman predictor)

The filtering problem discussed so far means the estimation of the current value of a random signal in the presence of additive noise. It is often required, particularly in control systems, to predict ahead if possible. Depending on how many steps of unit time ahead we want to predict, we distinguish one-step, two-step, or m-step prediction. The more steps we take, or the further in the future we want to look, the larger the prediction error will be. We deal here only with one-step prediction.

The signal model is again a first-order autoregressive process, described in section 7.3.1,

$$x(k) = ax(k-1) + w(k-1) \tag{7.53}$$

and the observation (or measurement) is affected by additive white noise, i.e.

$$y(k) = cx(k) + v(k) \tag{7.54}$$

We want the 'best' linear estimate of $x(k+1)$, i.e. the signal at time $k+1$, given the data and previous estimate at time k. We denote this one-step prediction estimate as $x(k+1\,|\,k)$. By 'best' we mean the predictor that minimizes the mean-square prediction error

$$\begin{aligned} p(k+1\,|\,k) &= E[e^2(k+1\,|\,k)] \\ &= E[x(k+1) - \hat{x}(k+1\,|\,k)]^2 \end{aligned} \tag{7.55}$$

This corresponds to the mean-square error

$$p(k) = E[x(k) - \hat{x}(k)]^2$$

in the filtering case. Strictly, a filtered estimate should be denoted as $\hat{x}(k\,|\,k)$.

For a one-step linear predictor, we choose the recursive form used earlier, i.e.

$$\hat{x}(k+1\,|\,k) = \alpha(k)\hat{x}(k\,|\,k-1) + \beta(k)y(k) \tag{7.56}$$

The parameters $\alpha(k)$ and $\beta(k)$ are determined from the minimization of the mean-square prediction error given by equation 7.55. Substituting equation 7.56 into equation 7.55 and differentiating, we obtain a set of orthogonal equations similar to those derived in the previous section

$$E[e(k+1\,|\,k)\hat{x}(k\,|\,k-1)] = 0 \tag{7.57}$$

$$E[e(k+1\,|\,k)y(k)] = 0 \tag{7.58}$$

The relationship between $\alpha(k)$ and $\beta(k)$,

$$\alpha(k) = a - c\beta(k) \tag{7.59}$$

is determined, from equation 7.57, in a similar way to the relationship between $a(k)$ and $b(k)$ derived in section 8 of the appendix for the filter case. Substituting this result into the prediction equation, we have

$$\hat{x}(k+1\,|\,k) = a\hat{x}(k\,|\,k-1) + \beta(k)[y(k) - c\hat{x}(k|k-1)] \tag{7.60}$$

The parameter $\beta(k)$ is determined, together with $p(k+1|k)$, from equations 7.58 and 7.55. Using a similar method to that used in section 9 of the appendix, we obtain

$$p(k+1|k) = \frac{a}{c}\sigma_v^2\beta(k) + \sigma_w^2 \tag{7.61}$$

where $$\beta(k) = \frac{acp(k|k-1)}{c^2p(k|k-1) + \sigma_v^2} \tag{7.62}$$

Equation 7.62 enables us to calculate $\beta(k)$ from the previous mean-square predictor error, and equation 7.61 then gives us the mean-square predictor error for $p(k+1|k)$.

As before, the optimum processor multiplies the previous estimate by a, and then adds a weighted correction term. Note that the correction term consists of the exact difference between the new data sample $y(k)$ and the previous prediction estimate $c\hat{x}(k|k-1)$. In the filtering problem considered in section 7.3.2, the correction term involved $y(k)$ minus a times the previous estimate, as seen by comparing equation 7.60 with equation 7.46.

Assuming that the random driving force in equation 7.53 is zero, the signal evolves according to the equation $x(k) = ax(k-1)$. Therefore, given an estimate $\hat{x}(k)$ at time k, it seems reasonable to predict the estimate at time $k+1$ as

$$\hat{x}(k+1\,|\,k) = a\hat{x}(k) \tag{7.63}$$

when no other information is available. It can be shown that this intuitive form of the estimate is also valid for the model driven by the noise $w(k-1)$. This is because, by hypothesis, the noise $w(k-1)$ is independent of the state at all times earlier than k (see Sorenson(26), pp. 226–8). Now applying equation 7.63 to equation 7.46, we have

$$\begin{aligned} a\hat{x}(k) &= \hat{x}(k+1|k) \\ &= a\hat{x}(k|k-1) + ab(k)[y(k) - c\hat{x}(k|k-1)] \end{aligned} \tag{7.64}$$

which is the same as the predictor equation 7.60, provided that $\beta(\mathrm{k}) = ab(k)$, i.e. the prediction gain and filtering gain are also related by the parameter a.

In addition, the mean-square estimation error $p(k)$ and the prediction error $p(k+1|k)$ are related as follows:

$$\begin{aligned} p(k+1|k) &= E[x(k+1) - \hat{x}(k+1|k)]^2 \\ &= E[ax(k) + w(k) - a\hat{x}(k)]^2 \\ &= E\{a[x(k) - \hat{x}(k)] + w(k)]\}^2 \end{aligned}$$

where we have used equations 7.53 and 7.63. Since $w(k)$ is not correlated with the error term $x(k) - \hat{x}(k)$, we have

$$p(k+1|k) = a^2 p(k) + \sigma_w^2 \tag{7.65}$$

Applying this to equation 7.61, with equation 7.62 substituted for $\beta(k)$, we obtain the mean-square estimation error $p(k)$ which is the same as $p(k)$ from equations 7.47 and 7.48, derived in section 7.3.2.

The optimum one-step prediction is shown in fig. 7.6, and optimum filtering and prediction simultaneously are shown in fig. 7.7. Solutions for the one-step predictor are given by equations 7.60 to 7.62, but we need these equations in a form suitable for direct

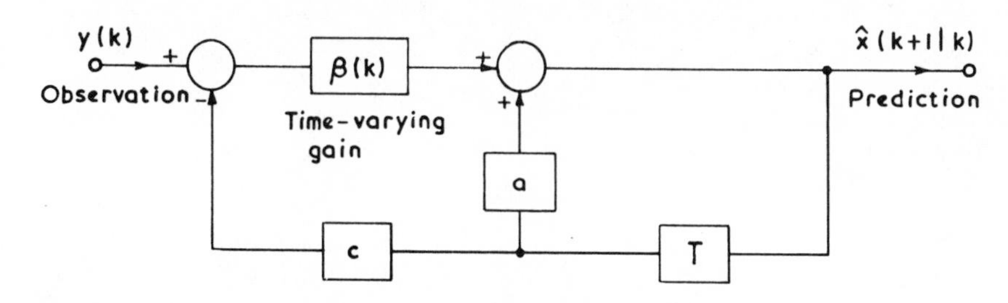

Fig. 7.6 Optimum recursive one-step predictor

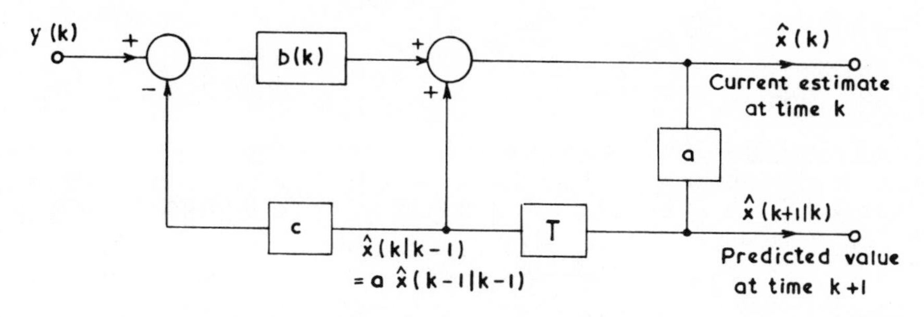

Fig. 7.7 Filtering and prediction simultaneously

transformation to vector signals. Equations 7.60 and 7.62 are already in this form, but equation 7.61 must be modified using equation 7.62 to eliminate σ_v^2.

The summarized set of equations is

Predictor equation:

$$\hat{x}(k+1|k) = a\hat{x}(k|k-1) + \beta(k)[y(k) - c\hat{x}(k|k-1)] \tag{7.66}$$

Predictor gain:

$$\beta(k) = acp(k|k-1)[c^2p(k|k-1) + \sigma_v^2]^{-1} \tag{7.67}$$

Prediction mean-square error:

$$p(k+1|k) = a^2p(k|k-1) - ac\beta(k)p(k|k-1) + \sigma_w^2 \tag{7.68}$$

We shall refer to the above set as the *scalar Kalman predictor* for the signal model

$$x(k+1) = ax(k) + w(k) \tag{7.69}$$

and the measurement or observation model

$$y(k) = cx(k) + v(k) \tag{7.70}$$

We return to these equations in section 8.4.

8

Optimum estimation of vector signals

8.0 Introduction

In this chapter we deal with vector or multidimensional signals. It is shown in section 8.1 how vector equations are formulated in the case of simultaneous estimation of a number of signals, or in the case of signals generated by higher-order systems. These vector equations result in matrix operations on vectors, so as discussed in section 8.2, the estimation problem for multidimensional systems is formulated in terms of vectors and matrices. These are of the same form as the scalar equations in chapter 7, so since there is an equivalence between scalar and matrix operations, all results in chapter 7 for scalar signals are transformed into vector and matrix equations in sections 8.3 and 8.4. To complete the multidimensional analysis we also extend the scalar Wiener filter of section 7.1 into a vector filter in section 8.5.

8.1 Signal and data vectors

We have dealt so far with scalar random signals generated by a first-order autoregressive process. We want to extend the same type of representation to broader classes of signals, and also to simultaneous estimation of several signals (for example, components of three-dimensional position and velocity vectors). It is shown below that these multidimensional signals are conveniently represented by vector notation, and in place of simple gain parameters we then have matrix operations on vectors. To illustrate the formation of vector equations we consider several examples (Schwartz & Shaw (21), p. 340).

Example 8.1

Assuming that we have q independent signals to be estimated or predicted *simultaneously*, we denote samples of these signals, at time k, as $x_1(k), x_2(k), \ldots, x_q(k)$. Assuming also that each one is generated by its own first-order autoregressive process, the αth signal sample, for example, is formed according to the equation

$$x_\alpha(k) = a_\alpha x_\alpha(k-1) + w_\alpha(k-1) \qquad \text{where } \alpha = 1, 2, \ldots, q \tag{8.1}$$

Each of the w_α processes is assumed to be white, zero-mean and independent of all others. We define the q-dimensional vectors made up of the q signals and q white noise driving processes as

$$\mathbf{x}(k) = \begin{bmatrix} x_1(k) \\ x_2(k) \\ \cdot \\ \cdot \\ \cdot \\ x_q(k) \end{bmatrix} \quad \text{and } \mathbf{w}(k) = \begin{bmatrix} w_1(k) \\ w_2(k) \\ \cdot \\ \cdot \\ \cdot \\ w_q(k) \end{bmatrix} \tag{8.2}$$

In terms of these defined vectors, the q equations 8.1 can be written as the *first-order vector equation*

$$\mathbf{x}(k) = A\,\mathbf{x}(k-1) + \mathbf{w}(k-1) \tag{8.3}$$

where $\mathbf{x}(k)$, $\mathbf{x}(k-1)$ and $\mathbf{w}(k-1)$ are $(q \times 1)$ column vectors and A is a $(q \times q)$ matrix, diagonal in this case, given by

$$A = \begin{bmatrix} a_1 & 0 & . & . & 0 \\ 0 & a_2 & . & . & 0 \\ \cdot & & & & \\ \cdot & & & & \\ \cdot & & & & \\ 0 & . & . & . & a_q \end{bmatrix} \tag{8.4}$$

Example 8.2

Let the signal $x(k)$ obey not a first-order but a second-order recursive difference equation

$$x(k) = ax(k-1) + bx(k-2) + w(k-1) \tag{8.5}$$

Such an equation can often arise in a real system, describing its dynamical behaviour, or it may fit a measured set of data better than the first-order difference equation.

To transform equation 8.5 into the first-order vector equation we define two components

$$x_1(k) = x(k) \quad \text{and} \quad x_2(k) = x(k-1) = x_1(k-1)$$

Now equation 8.5 can be written as two equations

$$\left.\begin{aligned} x_1(k) &= ax_1(k-1) + bx_2(k-1) + w(k-1) \\ x_2(k) &= x_1(k-1) \end{aligned}\right\} \tag{8.6}$$

where the first equation is equation 8.5 rewritten using defined components (or states), and the second equation represents the relationship between defined components.

Forming the two-dimensional vector

$$\mathbf{x}(k) = \begin{bmatrix} x_1(k) \\ x_2(k) \end{bmatrix}$$

we can combine the two equations 8.6 into the single vector equation

$$\underbrace{\begin{bmatrix} x_1(k) \\ x_2(k) \end{bmatrix}}_{\mathbf{x}(k)} = \underbrace{\begin{bmatrix} a & b \\ 1 & 0 \end{bmatrix}}_{A} \underbrace{\begin{bmatrix} x_1(k-1) \\ x_2(k-1) \end{bmatrix}}_{\mathbf{x}(k-1)} + \underbrace{\begin{bmatrix} w(k-1) \\ 0 \end{bmatrix}}_{\mathbf{w}(k-1)} \tag{8.7}$$

This is again in the form of the first-order vector equation.

Example 8.3
We consider a radar tracking problem, to be treated in more detail in chapter 9. Assume a vehicle being tracked is at range $R+\rho(k)$ at time k, and at range $R+\rho(k+1)$ at time $k+1$, T seconds later. We use T to represent the spacing between samples made one scan apart. The average range is denoted by R, and $\rho(k)$, $\rho(k+1)$ represent deviations from the average. We are interested in estimating these deviations, which are assumed to be statistically random with zero-mean value.

To a first approximation, if the vehicle is travelling at radial velocity $\dot{\rho}(k)$ and T is not too large

$$\rho(k+1)=\rho(k)+T\dot{\rho}(k) \tag{8.8}$$

which is the range equation, as given for range prediction in section 1.2.

Similarly, considering acceleration $u(k)$ we have

$$Tu(k)=\dot{\rho}(k+1)-\dot{\rho}(k) \tag{8.9}$$

which is the acceleration equation. Assuming that $u(k)$ is a zero-mean, stationary white noise process, the acceleration is, on average, zero and uncorrelated between intervals, i.e. $E[u(k+1)u(k)]=0$, but it has some known variance $E[u^2(k)]=\sigma_u^2$. Such accelerations might be caused by sudden wind gusts or short-term irregularities in engine thrust. The quantity $u_1(k)=Tu(k)$ is also a white noise process, and we have in place of equation 8.9

$$\dot{\rho}(k+1)=\dot{\rho}(k)+u_1(k) \tag{8.10}$$

We define now a two component signal vector $\mathbf{x}(k)$ with one component the range, $x_1(k)=\rho(k)$, and the other component the radial velocity, $x_2(k)=\dot{\rho}(k)$. Applying these to equations 8.8 and 8.10, we have

$$\begin{aligned} x_1(k+1) &= x_1(k)+Tx_2(k) \\ x_2(k+1) &= x_2(k)+u_1(k) \end{aligned}$$

or combining them into single vector equation, we obtain

$$\underbrace{\begin{bmatrix} x_1(k+1) \\ x_2(k+1) \end{bmatrix}}_{\mathbf{x}(k+1)} = \underbrace{\begin{bmatrix} 1 & T \\ 0 & 1 \end{bmatrix}}_{A} \underbrace{\begin{bmatrix} x_1(k) \\ x_2(k) \end{bmatrix}}_{\mathbf{x}(k)} + \underbrace{\begin{bmatrix} 0 \\ u_1(k) \end{bmatrix}}_{\mathbf{w}(k)} \tag{8.11}$$

which is again in the form of the first-order vector equation 8.3.

Equations with time-varying coefficients can be handled as well by defining a time-varying matrix $A(k, k-1)$, also known as the system transition matrix; see, for example, Sage & Melsa (27).

8.1.1 Data vector

Assume that in estimating the signal vector $\mathbf{x}(k)$ we made r *simultaneous noisy measurements at time* k. These measurement samples are labelled $y_1(k)$, $y_2(k)$, . . ., $y_r(k)$, so we have the following set of data.

$$
\begin{aligned}
y_1(k) &= c_1 x_1(k) + v_1(k) \\
y_2(k) &= c_2 x_2(k) + v_2(k) \\
&\vdots \\
y_r(k) &= c_r x_r(k) + v_r(k)
\end{aligned}
\tag{8.12}
$$

where $v_j(k)$ terms represent additive noise and $c_1, \ldots, c_r$ are some measurement parameters, similar to c introduced in equation 7.37. This set of equations can be put into vector form by defining r-component vectors $\mathbf{y}(k)$ and $\mathbf{v}(k)$. In terms of the previously defined q-component signal vector $\mathbf{x}(k)$, we then have the *data vector*:

$$\mathbf{y}(k) = C\,\mathbf{x}(k) + \mathbf{v}(k) \tag{8.13}$$

where $\mathbf{y}(k)$ and $\mathbf{v}(k)$ are $(r \times 1)$ column vectors, $\mathbf{x}(k)$ is a $(q \times 1)$ row vector, and C is an $(r \times q)$ observation matrix, which in this case, assuming $r < q$, is given by

$$
C = \overset{\xrightarrow{\quad q \quad}}{\left[\begin{array}{cccccccc}
c_1 & 0 & . & . & 0 & . & . & 0 \\
0 & c_2 & & & . & & & . \\
. & & & . & . & & & . \\
. & & & & . & & & . \\
0 & . & & & c_r & & & 0
\end{array}\right]} \; r\downarrow
\tag{8.14}
$$

Example 8.4
If $y_1(k) = 4\,x_1(k) + 5\,x_2(k) + v_1(k)$, and all the other coefficients in equation 8.12 are equal to unity, we have

$$
\begin{aligned}
y_1(k) &= 4\,x_1(k) + 5\,x_2(k) + 0\,x_3(k) + \ldots + v_1(k) \\
y_2(k) &= 0\,x_1(k) + x_2(k) + 0\,x_3(k) + \ldots + v_2(k) \\
&\vdots \\
y_r(k) &= 0\,x_1(k) + \ldots\ldots\ldots \qquad + x_r(k) + v_r(k)
\end{aligned}
$$

In this case the observation matrix is

$$
C = \left[\begin{array}{cccccccc}
4 & 5 & 0 & . & . & . & . & 0 \\
0 & 1 & 0 & & & & & . \\
0 & 0 & 1 & & & & & . \\
. & & & & & & & \\
. & & & & & & & \\
0 & . & . & . & 1 & . & . & 0
\end{array}\right]
$$

Example 8.5
For the previous radar tracking case, the signals to be estimated are the range $\rho(k)$, the

radial velocity $\dot{\rho}(k)$, the bearing (azimuth) $\theta(k)$, and the angular velocity $\dot{\theta}(k)$. Then $\mathbf{x}(k)$ is the four-component vector

$$\mathbf{x}(k) = \begin{bmatrix} \rho(k) \\ \dot{\rho}(k) \\ \theta(k) \\ \dot{\theta}(k) \end{bmatrix}$$

However, measurements of range and bearing only are made, in the presence of additive noise $v_1(k)$ and $v_2(k)$ respectively. In this case we have $q = 4$ and $r = 2$, and the velocities are then found in terms of these quantities, using equations such as 8.8. The matrix C is given here by

$$C = \begin{bmatrix} 1 & 0 & 0 & 0 \\ 0 & 0 & 1 & 0 \end{bmatrix}$$

The block diagram for the system and measurements in vector forms is the same as for the scalar case, figs 7.3 and 7.4 respectively. However, now the notation changes to vectors, and the system and observation parameters become matrices, as in figs 8.1 and 8.2. Here and in other figures shown later, vectors are denoted by underbars.

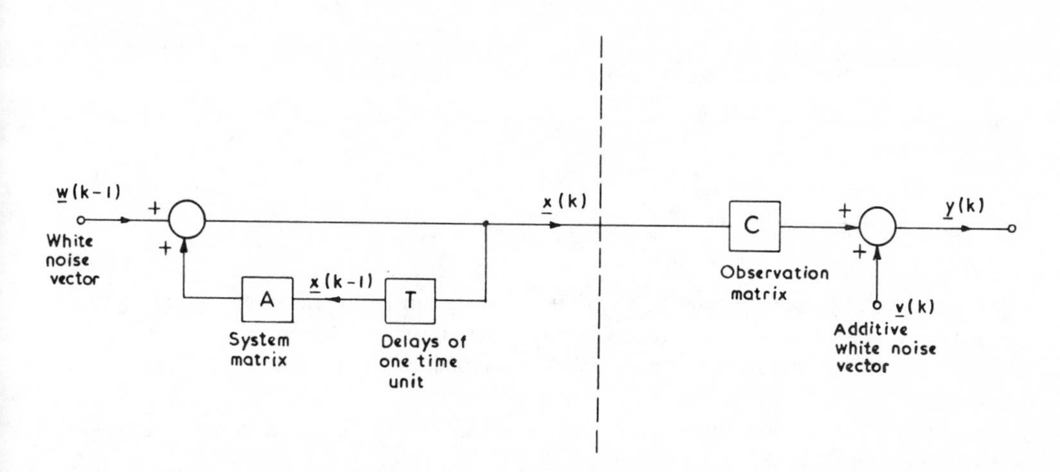

Fig. 8.1 System model

Fig. 8.2 Measurement (observation) model

8.2 Vector problem formulation

Returning to the basic problem, we have a signal vector $\mathbf{x}(k)$ obeying a known first-order vector dynamical equation

$$\mathbf{x}(k+1) = A\,\mathbf{x}(k) + \mathbf{w}(k) \tag{8.15}$$

to be extracted from a noisy measurement vector $\mathbf{y}(k)$

$$\mathbf{y}(k) = C\,\mathbf{x}(k) + \mathbf{v}(k) \tag{8.16}$$

These two vector equations are obtained as discussed in section 8.1.

The problem is how to form $\hat{\mathbf{x}}(k)$, the 'best' linear estimate (filtered value) of $\mathbf{x}(k)$, and how to form $\hat{\mathbf{x}}(k \mid k-1)$, the 'best' predicted value. By 'best' we now mean estimators that minimize the mean-square error of each signal component simultaneously. For example, in the filtering operation each mean-square error

$$E[x_\alpha(k) - \hat{x}_\alpha(k)]^2 \quad \text{where } \alpha = 1, 2, \ldots, q \tag{8.17}$$

is to be minimized.

The problem is formally the same as that stated previously in dealing with single time-varying signals obeying a first-order dynamical equation. In the multidimensional case, or the more complex signal case, we could reformulate all the equations in vector form and apply a matrix minimization procedure to obtain the optimum solutions. This procedure would be very similar to the ones used for the single signal case, but now repeated in vector form (see, for example, Mendel (22), p. 64). However, since we already have the solutions for the one-dimensional (scalar) cases we can extend them to the multidimensional (vector) systems, using the equivalence of scalar and matrix operations given by Scovell (28) and in Table 8.1, in which the superscript T stands for transpose of a matrix, and -1 for the inverse of a matrix.

Table 8.1 Transformation of scalar to matrix

Scalar →	*Matrix*
$a+b$	$A+B$
ab	AB
a^2b	ABA^{T}
$\dfrac{1}{a+b}$	$(A+B)^{-1}$

We have already seen, in section 8.1, that in transition from the single signal to vector signal, the system parameter a changed into the system matrix A, and the data coefficient c changed into the observation matrix C. We now consider the transition of other relevant quantities.

The transition from the observation noise variance to the observation noise covariance matrix (common variance of a number of signals) is written as

$$\sigma_v^2 = \sigma_{v1,1}^2 = E[v_1^2(k)] \rightarrow R(k) = E[\mathbf{v}(k)\mathbf{v}^{\mathrm{T}}(k)] \tag{8.18}$$

where we have used the third entry of Table 8.1, with $b = 1$. For example, for two signals, we have

$$R(k) = \begin{bmatrix} E[v_1^2(k)] & E[v_1(k)\, v_2(k)] \\ E[v_2(k)\, v_1(k)] & E[v_2^2(k)] \end{bmatrix}$$

$$= \begin{bmatrix} \sigma_{v1,1}^2 & \sigma_{v1,2}^2 \\ \sigma_{v2,1}^2 & \sigma_{v2,2}^2 \end{bmatrix}$$

Similarly, for the system noise, we have

$$\sigma_w^2 = \sigma_{w1,1}^2 = E[w_1^2(k)] \rightarrow Q(k) = E[\mathbf{w}(k)\mathbf{w}^{\mathrm{T}}(k)] \tag{8.19}$$

where $Q(k)$ represents the system noise covariance matrix. If there is no correlation between noise processes, the off-diagonal terms are zero.

The mean-square error for the single signal changes into the error covariance matrix

$$p(k) = p_{1,1}(k) = E[e_1^2(k)] \rightarrow P(k) = [\mathbf{e}(k)\mathbf{e}^{\mathrm{T}}(k)] \tag{8.20}$$

For two signals, we have

$$P(k) = \begin{bmatrix} E[e_1^2(k)] & E[e_1(k)\,e_2(k)] \\ E[e_2(k)\,e_1(k)] & E[e^2(k)] \end{bmatrix} = \begin{bmatrix} p_{1,1}(k) & p_{1,2}(k) \\ p_{2,1}(k) & p_{2,2}(k) \end{bmatrix} \tag{8.21}$$

Note that the diagonal terms are the individual mean-square errors as formulated by equation 8.17.

8.3 Vector Kalman filter

We are now in a position to transform the scalar Kalman filter algorithm, given by equations 7.49 to 7.52, into the corresponding vector Kalman filter. With reference to these equations, and discussions in sections 8.1 and 8.2, we can write directly the vector and matrix equations, tabulated below.

Estimator:

$$\hat{\mathbf{x}}(k) = A\,\hat{\mathbf{x}}(k-1) + K(k)[\mathbf{y}(k) - CA\,\hat{\mathbf{x}}(k-1)] \tag{8.22}$$

Filter gain:

$$K(k) = P_1(k)C^{\mathrm{T}}[CP_1(k)C^{\mathrm{T}} + R(k)]^{-1} \tag{8.23}$$

where $$P_1(k) = AP(k-1)\,A^{\mathrm{T}} + Q(k-1) \tag{8.24}$$

Error covariance matrix:

$$P(k) = P_1(k) - K(k)C(k)\,P_1(k) \tag{8.25}$$

The above equations constitute the *vector Kalman filter* for the model described by the state equations

$$\mathbf{x}(k) = A\,\mathbf{x}(k-1) + \mathbf{w}(k-1) \tag{8.26}$$

$$\mathbf{y}(k) = C\,\mathbf{x}(k) + \mathbf{v}(k) \tag{8.27}$$

introduced and discussed in section 8.1(equations 8.3 and 8.13).

According to the scalar-matrix equivalence Table 8.1, it would have been more correct to write $B(k)$ in place of $b(k)$. We have used $K(k)$ instead, since this is a commonly used notation for the gain matrix in the Kalman filter. Other quantities have been arranged, within the rules of Table 8.1, to obtain the standard form of Kalman equations used, for example, by Sorenson (26) and Jazwinski (29). It will be found that some authors use the notation $P(k \mid k-1)$ and $P(k \mid k)$ in place of $P_1(k)$ and $P(k)$ respectively. The quantity $P(k \mid k-1)$ is then referred to as the *predicted covariance matrix*. The notation $P_1(k)$ and $P(k)$, based on a similar one used by Sorenson (26), has been chosen here for simplicity. In equation 8.24, we have

$Q(k-1)$ since σ_w^2 in fact represents $E[w^2(k-1)]$, as shown in section 9 of the appendix. As mentioned earlier in section 8.1, for time-varying systems and time-varying observations, matrices A and C are functions of time, i.e. we have $A(k, k-1)$ and $C(k)$ respectively.

To check that equations 8.22 to 8.25 are written correctly, we write the corresponding dimensional equations:

$$q \times 1 = (q \times q)(q \times 1) + (q \times r)[(r \times 1) - (r \times q)(q \times q)(q \times 1)] \tag{8.22'}$$

$$q \times r = (q \times q)(q \times r)[(r \times q)(q \times q)(q \times r) + (r \times r)]^{-1} \tag{8.23'}$$

Similar equations can be written for equations 8.24 and 8.25. This is in general a very useful method of checking whether vector or matrix equations are written correctly.

Eliminating $K(k)$ from equation 8.25 by means of equation 8.23, changing $P(k)$ to $P(k-1)$, and substituting it into equation 8.24, we obtain a nonlinear difference equation. This equation is known as the *Ricatti* difference equation whose solution in a closed form is generally complex. However, if a numerical procedure of the Kalman algorithm (i.e. equations 8.22 to 8.25) is carried out, there is no need to have a closed form solution for $P_1(k)$.

One of the most significant features of the Kalman filter is its recursive form property that makes it extremely useful in processing measurements to obtain the optimal estimate, utilizing a digital computer. The measurements may be processed as they occur, and it is not necessary to store any measurement data. Only $\hat{\mathbf{x}}(k-1)$ need to be stored in proceeding from time $(k-1)$ to time k. However, the algorithm does require storage of the time histories of the matrices $A(k)$ and $C(k)$ if they are time-varying, $Q(k-1)$, and $R(k)$ for all $k = 1, 2 \ldots$. These fall into two types: first, the physical model must be defined (A, C), and second, the statistics of the random processes must be known (Q, R).

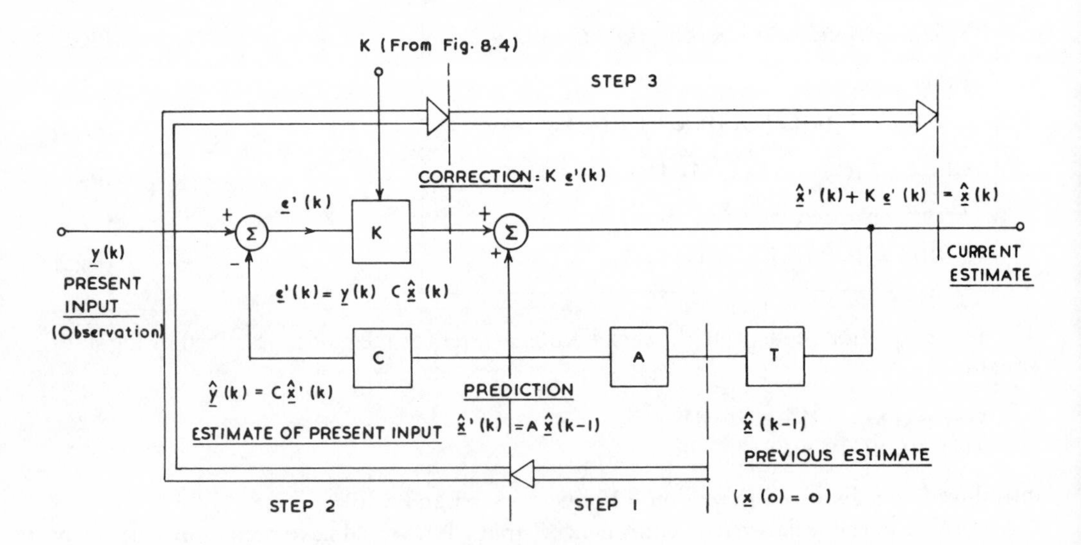

Fig. 8.3 Computational steps in Kalman filter

The information flow in the filter can be discussed very simply by considering the block diagram fig. 8.3, which is a representation of equation 8.22. Let us suppose that $\hat{\mathbf{x}}(k-1)$ is known for some k and that we seek to determine $\hat{\mathbf{x}}(k)$, given $\mathbf{y}(k)$. The computational cycle would proceed as follows:

(i) the estimate $\hat{\mathbf{x}}(k-1)$ is 'propagated forward' by premultiplying it by the system matrix A. This gives the predicted estimate $\hat{\mathbf{x}}(k \mid k-1)$, denoted as $\hat{\mathbf{x}}'(\mathbf{k})$;
(ii) $\hat{\mathbf{x}}'(k)$ is premultiplied by C giving $\hat{\mathbf{y}}(k)$ which is subtracted from the actual measurement $\mathbf{y}(k)$ to obtain the measured residual (or error) $\mathbf{e}'(k)$;
(iii) the residual is premultiplied by the matrix $K(k)$ and the result is added to $\hat{\mathbf{x}}'(k)$ to give $\hat{\mathbf{x}}(k)$;
(iv) $\hat{\mathbf{x}}(k)$ is stored until the next measurement is made, at which time the cycle is repeated.

The filter operates in a 'predict-correct' fashion, i.e. the 'correction' term $K(k)\,\mathbf{e}'(k)$ is added to the predicted estimate $\hat{\mathbf{x}}'(k)$ to determine the filtered estimate. The correction term involves the Kalman gain matrix $K(k)$. A similar discussion is given in section 7.4 in the paragraph below equation 7.62. In order to initiate filtering, we start with $\hat{\mathbf{x}}(0) = 0$, and we see immediately that $\hat{\mathbf{x}}(1) = K(1)\,\mathbf{y}(1)$. Then $\hat{\mathbf{x}}(2), \hat{\mathbf{x}}(3), \ldots$ follow recursively, as discussed in the above four steps.

It is interesting to note that the optimal filter shown in fig. 8.3 consists of the *model of the dynamic process*, which performs the function of prediction, and a feedback correction scheme in which the product of gain and the residual term is applied to the model as a forcing function (compare this system with fig. 8.1). The same comment applies to the scalar Kalman filter shown in fig. 7.5 and the model in fig. 7.3.

The gain matrix $K(k)$ can be calculated before estimation is carried out since it does not depend at all on the measurements, as already shown in example 7.5. This approach requires storing the calculated vectors for each recursion and feeding them out as needed. However, in the subroutine algorithm given by equations 8.23 to 8.25, in which gains are updated recursively as the estimation proceeds, there is no need to store all gain values: the previous value is the only one required. In this case the subroutine computational diagram is shown in fig. 8.4, and the computational cycle would proceed as follows:

(i) given $P(k-1)$, $Q(k-1)$, $A(k, k-1)$, then $P_1(k)$ is computed using equation 8.24;
(ii) $P_1(k)$, $C(k)$ and $R(k)$ are substituted into equation 8.23 to obtain $K(k)$, which is used in step three of the filter computation given earlier;
(iii) $P_1(k)$, $K(k)$ and $C(k)$ are substituted into equation 8.25 to determine $P(k)$, which is stored until the time of the next measurement, when the cycle is repeated.

The matrix inverse which must be computed in equation 8.23 is generally no problem. The matrix to be inverted is $(r \times r)$, as shown in equation 8.23′, where r is the number of elements in the measurement vector (see equation 8.13). In most systems r is kept small to avoid the high cost of complex instrumentation. Consequently, it is not unusual to encounter systems with 12 to 15 state variables (in vector $\mathbf{x}$), but only two to three measurement variables (vector $\mathbf{y}$). If r is large, it will also be computationally costly to perform the matrix inversion, and the alternative then is to apply the matrix inversion lemma, as given by Mendel (22), p. 96.

Equations 8.23 to 8.25 define the algorithm for the recursive computation of the optimal filter gain matrix $K(k)$. At the same time, we obtain values for $P_1(k)$ and $P(k)$, i.e. the variances of the components of the prediction and filtering errors respectively.

It is useful for calculations and also for better understanding to group the results for the Kalman filter, equations 8.22 to 8.25, into the following two stages:

(i) *Prediction*

$$\hat{\mathbf{x}}(k \mid k-1) = A\,\hat{\mathbf{x}}(k-1 \mid k-1) \tag{8.28}$$

$$P(k \mid k-1) = A\,P(k-1 \mid k-1)A^{\mathrm{T}} + Q(k-1) \tag{8.29}$$

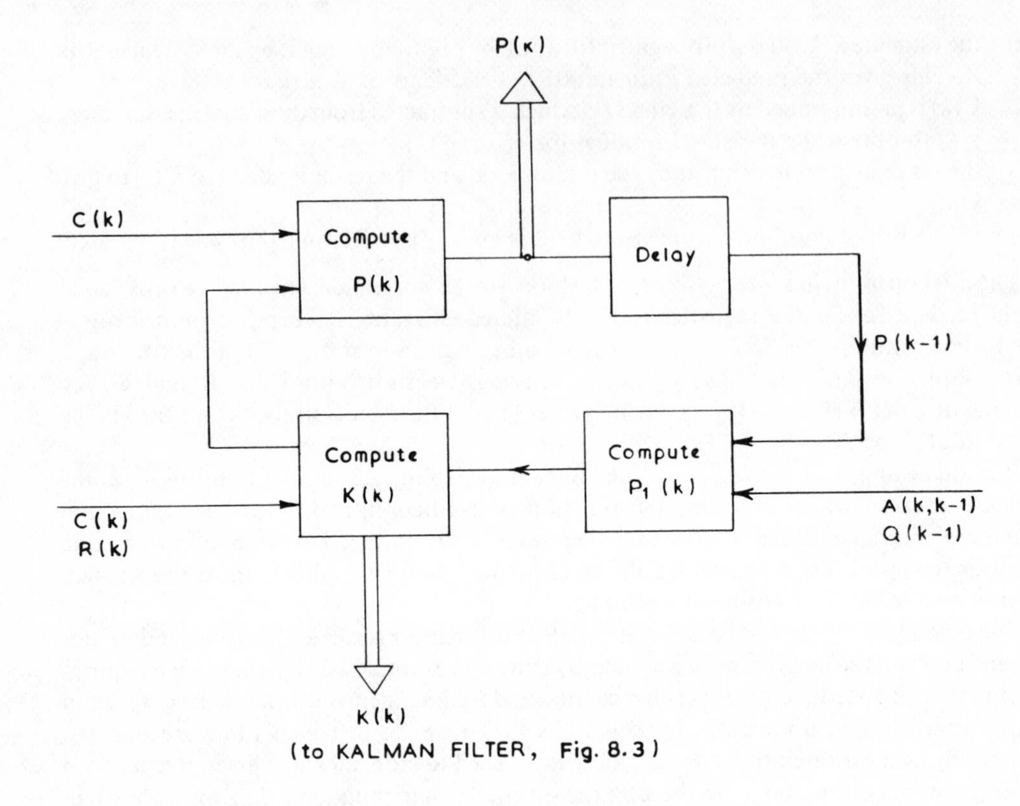

Fig. 8.4 Subroutine calculations for Kalman filter

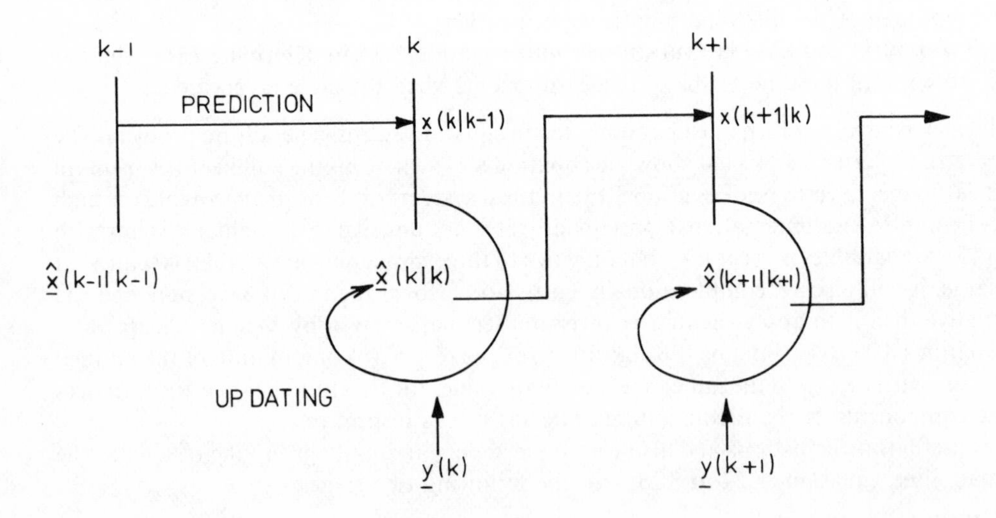

Fig. 8.5 Illustration for two stage Kalman computational cycle

(ii) *Updating (or correction)*

$$\hat{\mathbf{x}}(k\,|\,k) = \hat{\mathbf{x}}(k\,|\,k-1) + K[\mathbf{y}(k) - C\,\hat{\mathbf{x}}(k\,|\,k-1)] \tag{8.30}$$

$$P(k\,|\,k) = P(k\,|\,k-1) - K\,C\,P(k\,|\,k-1) \tag{8.31}$$

$$K = P(k\,|\,k-1)C^{\mathrm{T}}[C\,P(k\,|\,k-1)C^{\mathrm{T}} + R(k)]^{-1} \tag{8.32}$$

where $P(k\,|\,k-1)$ and $P(k\,|\,k)$ correspond to $P_1(k)$ and $P(k)$ respectively in the Kalman filter given by equations 8.23 to 8.25. The first stage is prediction based on the state equation 8.26, and the second stage is the updating or correction based on the measurement equation 8.27; these stages are also illustrated in fig. 8.5. This presentation of the Kalman filter is in a form particularly suitable for computations, but the earlier form given by equations 8.22 to 8.25 is more suitable as an introduction.

8.4 Vector Kalman predictor

We already have the scalar Kalman predictor algorithm, equations 7.66 to 7.68, arranged in a form suitable for transforation into vector equations. Using the same kind of reasoning as in sections 8.1 to 8.3, we obtain the following vector and matrix set of equations.

Predictor equation:

$$\hat{\mathbf{x}}(k+1\,|\,k) = A\,\hat{\mathbf{x}}(k\,|\,k-1) + G(k)[\mathbf{y}(k) - C\,\hat{\mathbf{x}}(k\,|\,k-1)] \tag{8.33}$$

Predictor gain:

$$G(k) = A\,P(k\,|\,k-1)C^{\mathrm{T}}[C\,P(k\,|\,k-1)C^{\mathrm{T}} + R(k)]^{-1} \tag{8.34}$$

Prediction mean-square error:

$$P(k+1\,|\,k) = [A - G(k)C]P(k\,|\,k-1)A^{\mathrm{T}} + Q(k) \tag{8.35}$$

These equations represent the *vector Kalman predictor* for the model described earlier by equations 8.26 and 8.27. We have introduced here $G(k)$ as the predictor gain matrix with its scalar value denoted in section 7.4 as $\beta(k)$. Other quantities are the same as the ones used in the previous section.

As in the scalar case given by equation 7.63, and by Sorenson (26), we have again the following interesting connection between estimated and predicted signal vectors

$$\hat{\mathbf{x}}(k+1\,|\,k) = A\,\hat{\mathbf{x}}(k) \tag{8.36}$$

This means that given the filtered signal $\hat{\mathbf{x}}(k)$, the best estimate of the signal one step in the future ignores noise and assumes that the signal dynamics matrix A operates only on the estimate. Multiplying the estimator (filter) equation 8.22 by A and using equation 8.31, we obtain the predictor equation 8.33, with $G(k) = AK(k)$. This is the relationship corresponding to the scalar relationship $\beta(k) = ab(k)$ discussed in section 7.4 below equation 7.64. Since in the prediction case we are interested in the prediction error covariance, the three matrix equations 8.23 to 8.25 reduce to two equations 8.34 and 8.35. The prediction error covariance matrix given by equation 8.35 is obtained after substitution of equation 8.25 into equation 8.24 and expressing $P_1(k)$ as $P(k|k-1)$.

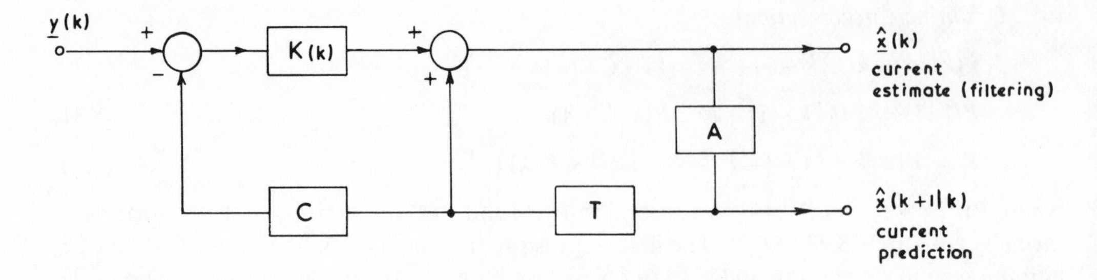

Fig. 8.6 Simultaneous filtering and prediction of vector signals

The simple link between filtering and prediction, given by equation 8.36, enables the one-step vector prediction to be obtained from the Kalman filter, as shown in fig. 8.6. This is, in fact, the vector case corresponding to fig. 7.7. Note that the A box and the time-delay element are interchanged in this version of the Kalman filter.

8.5 Vector Wiener filter

To complete the extension of scalar results to vector signals we develop in this section Wiener filter for vector signals. We start with the scalar result given by equation 7.4 rewritten here as

$$\sum_{i=1}^{m} h(i)E[y(i)y(j)] = E[xy(j)] \qquad \text{for } j = 1, 2, \ldots, m \tag{8.37}$$

In this equation signal x is a constant, and we operate on m data $y(1), y(2), \ldots, y(m)$. We want to extend this result to a time-varying signal denoted as $x(k)$, for $k = 1, 2, \ldots, m$. This means that now data samples will be time-varying not only due to additive noise $v(k)$, but due to $x(k)$ time changes. It seems logical then to assume that for equation 8.37 to hold in this case, the filter coefficients should become variable with k. Introducing k via $x(k)$ on the right-hand-side of equation 8.37, and via $h(k, i)$ on the left-hand-side, we have the *Wiener filter for time-varying signal* given by

$$\sum_{i=1}^{m} h(k, i)E[y(i)y(j)] = E[x(k)y(j)] \qquad \text{for } j, k = 1, 2, \ldots, m \tag{8.38}$$

To derive the vector equation we first consider the right-hand-side. Expanding it over k and j, and neglecting the operator E, we obtain the following array:

$$\xrightarrow{\quad j \quad}$$

$$k \downarrow \begin{bmatrix} x(1)y(1) & \ldots & x(1)y(j) & \ldots & x(1)y(m) \\ \vdots & & & & \\ x(k)y(1) & \ldots & x(k)y(j) & \ldots & x(k)y(m) \\ \vdots & & & & \\ x(m)y(1) & \ldots & x(m)y(j) & \ldots & x(m)y(m) \end{bmatrix}$$

This can be expressed as vector product $\mathbf{x}\mathbf{y}^{\mathrm{T}}$, where the vectors are defined as

$$\mathbf{x} = \begin{bmatrix} x(1) \\ \vdots \\ x(k) \\ \vdots \\ x(m) \end{bmatrix} \quad \text{and } \mathbf{y}^{\mathrm{T}} = [y(1) \ldots y(j) \ldots y(m)] \tag{8.39}$$

where $\mathbf{x}$ is an $(m \times 1)$ column vector and $\mathbf{y}^{\mathrm{T}}$ is a $(1 \times m)$ row vector. Therefore, the right-hand-side of equation 8.38 can be written as

$$E(\mathbf{x}\mathbf{y}^{\mathrm{T}}) \tag{8.40}$$

which is the cross-covariance $(m \times m)$ matrix of the signal and data samples.

Consider now a (k, j) element on the left-hand-side of equation 8.38, which for fixed (k, j) can be written as

$$\sum_{i=1}^{m} h(k, i) E[y(i)\, y(j)] = h(k, 1) E[y(1)\, y(j)] + h(k, 2) E[y(2)\, y(j)] + \ldots$$
$$+ h(k, m) E[y(m)\, y(j)]$$

In vector notation we can express the above as $\mathbf{h}^{\mathrm{T}}(k)\mathbf{p}_y(j)$, where

$$\mathbf{h}^{\mathrm{T}}(k) = [h(k, 1), h(k, 2), \ldots, h(k, m)] \tag{8.41}$$

$$\text{and} \quad \mathbf{p}_y(j) = \begin{bmatrix} p_y(1) \\ p_y(2) \\ \vdots \\ p_y(m) \end{bmatrix} = \begin{bmatrix} E[y(1)\, y(j)] \\ E[y(2)\, y(j)] \\ \vdots \\ E[y(m)\, y(j)] \end{bmatrix} \tag{8.42}$$

Expanding $\mathbf{h}^{\mathrm{T}}(k)\mathbf{p}_y(j)$ over k and j we obtain the following array

$$\xrightarrow{\quad j \quad}$$

$$k \downarrow \begin{bmatrix} \mathbf{h}^{\mathrm{T}}(1)\, \mathbf{p}_y(1) & \ldots & \mathbf{h}^{\mathrm{T}}(1)\, \mathbf{p}_y(j) & \ldots & \mathbf{h}^{\mathrm{T}}(1)\, \mathbf{p}_y(m) \\ \vdots & & & & \\ \mathbf{h}^{\mathrm{T}}(k)\, \mathbf{p}_y(1) & \ldots & \mathbf{h}^{\mathrm{T}}(k)\, \mathbf{p}_y(j) & \ldots & \mathbf{h}^{\mathrm{T}}(k)\, \mathbf{p}_y(m) \\ \vdots & & & & \\ \mathbf{h}^{\mathrm{T}}(m)\, \mathbf{p}_y(1) & \ldots & \mathbf{h}^{\mathrm{T}}(m)\, \mathbf{p}_y(j) & \ldots & \mathbf{h}^{\mathrm{T}}(m)\, \mathbf{p}_y(m) \end{bmatrix}$$

This can be expressed as the following matrix product

$$\underbrace{\begin{bmatrix} \mathbf{h}^{\mathrm{T}}(1) \\ \vdots \\ \mathbf{h}^{\mathrm{T}}(k) \\ \vdots \\ \mathbf{h}^{\mathrm{T}}(m) \end{bmatrix}}_{H} \underbrace{[\mathbf{p}_y(1) \ldots \mathbf{p}_y(j) \ldots \mathbf{p}_y(m)]}_{E(\mathbf{y}\mathbf{y}^{\mathrm{T}})} \tag{8.43}$$

Therefore, the Wiener filter for a time-varying signal, given by equation 8.38, can be written in matrix form as

$$HE(\mathbf{y}\mathbf{y}^{\mathrm{T}}) = E(\mathbf{x}\mathbf{y}^{\mathrm{T}}) \tag{8.44}$$

where $E(\mathbf{x}\ \mathbf{y}^{\mathrm{T}})$, $E(\mathbf{y}\ \mathbf{y}^{\mathrm{T}})$ are $(m \times m)$ matrices, and H is the following $(m \times m)$ matrix of filter coefficients:

$$H = \begin{bmatrix} h(1,1) & \dots & h(1,j) & \dots & h(1,m) \\ \vdots & & & & \\ h(k,1) & \dots & h(k,j) & \dots & h(k,m) \\ \vdots & & & & \\ h(m,1) & \dots & h(m,j) & \dots & h(m,m) \end{bmatrix} \tag{8.45}$$

In a multidimensional system we have, at time k, perhaps q signals $x_1(k), x_2(k), \dots, x_q(k)$, and r observation (data) values $y_1(k), y_2(k), \dots, y_r(k)$. It can be shown that in this case, equation 8.38 becomes the following *Wiener filter vector equation*

$$\sum_{i=1}^{m} H(k,i)\,E[\mathbf{y}(i)\mathbf{y}^{\mathrm{T}}(j)] = E[\mathbf{x}(k)\mathbf{y}^{\mathrm{T}}(j)] \qquad \text{for } j, k = 1, 2, \dots, m \tag{8.46}$$

where $H(k, i)$ is the following $(q \times r)$ matrix

$$H(k,i) = \begin{bmatrix} h_{1,1}(k,i) & \dots & h_{1,r}(k,i) \\ & \vdots & \\ h_{q,1}(k,i) & \dots & h_{q,r}(k,i) \end{bmatrix} \tag{8.47}$$

and $\mathbf{x}(k)$, $\mathbf{y}(k)$ are vectors

$$\mathbf{x}(k) = \begin{bmatrix} x_1(k) \\ \vdots \\ x_q(k) \end{bmatrix} \qquad \mathbf{y}(k) = \begin{bmatrix} y_1(k) \\ \vdots \\ y_r(k) \end{bmatrix} \tag{8.48}$$

Equation 8.46 can be derived from the one-dimensional equation 8.38, in a similar way to the vector formulation of equation 8.38. For this purpose we deal with $x_\alpha(k)$ and $y_\beta(k)$, where α and β refer to signals and data at time k (see, for example, section 8.1). In the derivation of equation 8.44 we expanded expressions over $j, k = 1, 2, \dots, m$, while in this case we expand over $\alpha = 1, \dots, q$, and $\beta = 1, \dots, r$.

To link the multidimensional Wiener filter with the scalar Wiener filter of section 7.1, we write equations corresponding to 7.8, 7.1 and 7.10. First, to obtain the equation analogous to 7.8, we rewrite equation 8.46 as

$$\sum_{i=1}^{m} H(k, i)\, P_y(i, j) = P_{xy}(k, j) \tag{8.49}$$

where $j, k = 1, \ldots, m$. $P_y(i, j)$ is the $(r \times r)$ autocorrelation matrix of $\mathbf{y}(i)$ and $\mathbf{y}(j)$, and $P_{xy}(k, j)$ is the $(q \times r)$ cross-correlation matrix of $\mathbf{x}(k)$ and $\mathbf{y}(k)$. Expanding equation 8.49 over $i, j, k = 1, \ldots, m$, we have

$$P_y = \begin{bmatrix} P_y(1, 1) & \ldots & P_y(1, m) \\ \vdots & & \\ P_y(m, 1) & \ldots & P_y(m, m) \end{bmatrix} \tag{8.50}$$

$$P_{xy} = \begin{bmatrix} P_{xy}(1, 1) & \ldots & P_{xy}(1, m) \\ \vdots & & \\ P_{xy}(m, 1) & \ldots & P_{xy}(m, m) \end{bmatrix} \tag{8.51}$$

Similarly, $H(k, i)$ produces the $(qm \times rm)$ matrix

$$H = \begin{bmatrix} H(1, 1) & \ldots & H(1, m) \\ \vdots & & \\ H(m, 1) & \ldots & H(m, m) \end{bmatrix} \tag{8.52}$$

where each element is a $(q \times r)$ matrix.
Therefore equation 8.49 can be written in the compact form as

$$H\, P_y = P_{xy}$$

with the solution

$$H = P_{xy}\, P_y^{-1} \tag{8.53}$$

This is, theoretically, a straightforward solution, but required matrix inversion becomes computationally impractical when qm, qr are large, as discussed by Sorenson (25).

The estimator equation for the multidimensional case, analogous to equation 7.1, is given by

$$\hat{\mathbf{x}}(k) = \sum_{i=1}^{m} H(k, i)\, \mathbf{y}(i) \qquad \text{for } i, k = 1, \ldots, m. \tag{8.54}$$

This equation follows from the convolution, see for example Mendel (22), p. 15.

The error covariance matrix, corresponding to scalar equation 7.10, can be shown to be

$$P(k) = P_x(k, k) - \sum_{i=1}^{m} H(k, i)\, P_{yx}(i, k) \tag{8.55}$$

where $P_x(k, k) = E[\mathbf{x}(k)\mathbf{x}^{\mathrm{T}}(k)]$ and $P_{yx}(i, k) = E[\mathbf{y}(i)\mathbf{x}^{\mathrm{T}}(k)]$.

The *vector* or *multidimensional Wiener filter* is then given by equations 8.49, 8.54 and 8.55.

To solve it we require the correlation matrices P_{xy} and P_y.

For the case of a linear measurement model given by equation 8.13, we have

$$\begin{aligned} P_{xy}(k,j) &= E\{\mathbf{x}(k)[C\,\mathbf{x}(j)+\mathbf{v}(j)]^{\mathrm{T}}\} \\ &= P_x(k,j)\,C^{\mathrm{T}} \end{aligned} \tag{8.56}$$

and $$\begin{aligned} P_y(i,j) &= E\{[C\,\mathbf{x}(i)+\mathbf{v}(i)][C\,\mathbf{x}(j)+\mathbf{v}(j)]^{\mathrm{T}}\} \\ &= C\,P_x(i,j)C^{\mathrm{T}} + R(i)\,\delta(i,j) \end{aligned} \tag{8.57}$$

assuming signal $\mathbf{x}$ and measurement noise $\mathbf{v}$ to be independent. In the above, $R(i)$ is the observation noise covariance discussed earlier, equation 8.18, and $\delta(i,j)$ is the Kronecker delta used in example 7.1. In fact, the equations 8.56 and 8.57 correspond respectively to equations 7.14 and 7.13 in example 7.1. From the above two equations, we see that to solve the problem it is sufficient, in this case, to know P_x and R. This and other kinds of modification of Wiener filter can be found, for example in Liebelt (30), chapter 5. If the signal generation equation is also known, further modifications are possible, as illustrated in example 9.5.

9
Examples

9.0 Introduction

The notation in part 2 such as x, $\hat{x}$, y, P, Q, R, K is the same as that generally used in the literature on Kalman filtering, but the notation used in state equations is different. We are using the discrete-time state equations written as

$$\begin{aligned} \mathbf{x}(k) &= A\mathbf{x}(k-1) + \mathbf{w}(k-1) \\ \mathbf{y}(k) &= C\mathbf{x}(k) + \mathbf{v}(k) \end{aligned} \tag{9.1}$$

where A and C, if time-varying, would be denoted respectively as $A(k, k-1)$ and $C(k)$. This notation is analogous to the continuous-time state equations written as

$$\begin{aligned} \dot{\mathbf{x}}(t) &= A(t)\mathbf{x}(t) + \mathbf{w}(t) \\ \mathbf{y}(t) &= C(t)\mathbf{x}(t) \end{aligned} \tag{9.2}$$

The alternative notation to the one used in equations 9.1 is

$$\begin{aligned} \mathbf{x}(k) &= \Phi(k, k-1)\,\mathbf{x}(k-1) + \mathbf{w}(k-1) \\ \mathbf{y}(k) &= H(k)\mathbf{x}(k) + \mathbf{v}(k) \end{aligned} \tag{9.3}$$

where $\Phi(k, k-1)$, sometimes denoted as $F(k)$, is the state transition matrix corresponding to $A(k, k-1)$ in equation 9.1, and $H(k)$ is the observation matrix corresponding to $C(k)$ in equation 9.1. The notation of equation 9.3 will be found in most literature on Kalman filtering (or general estimation theory), for example in references (21) to (28). However, it is interesting to note that in some text books (for example Freeman (6), p. 19 and Eykhoff (23), p. 432), A, B, C notation is used for the discrete-time systems in the same way as in this book, where the additional letter B is the matrix relating the control input to the state vector, which has not been used in this book except later in example 9.4.

There are six examples in this chapter. The first three are standard ones illustrating the basic manipulation techniques for Wiener and Kalman filters. The others are more advanced, dealing with falling body estimation using Kalman and then Wiener filter, and Kalman filter formulation for radar tracking.

9.1 Scalar Wiener filter

Estimate the random amplitude x of a sinusoidal waveform of known frequency ω, in the presence of additive noise $v(t)$, using the scalar Wiener filter derived in section 7.1. The measurement is represented as

$$y(t) = x\cos\omega t + v(t) \tag{9.4}$$

where the additive noise $v(t)$ accounts for both receiver noise and inaccuracies of the instruments used. Let $y(t)$ be sampled at $\omega t = 0$ and $\omega t = \pi/4$, giving us measured data $y(1) = y(\omega t = 0)$ and $y(2) = y(\omega t = \pi/4)$. Assume that $E(x) = 0$, $E(x^2) = \sigma_x^2$, $E[v(1)v(2)] = 0$, $E[v(1)]^2 = E[v(2)]^2 = \sigma_v^2$, and also that x is independent of v.

The linear estimator for this case is given by equation 7.1 with optimum weights expressed by equation 7.8 or 7.4. We have two measurement samples $y(1) = x + v_1$ and $y(2) = (x/\sqrt{2}) + v_2$. Therefore, expanding equation 7.8 for $i = j = 1, 2$, we have

$$\begin{aligned} h(1)\,p_y(1,1) + h(2)\,p_y(2,1) &= p_{xy}(1) \\ h(2)\,p_y(1,2) + h(2)\,p_y(2,2) &= p_{xy}(2) \end{aligned} \tag{9.5}$$

where $p_y(i,j)$ and $p_{xy}(j)$ are given by equations 7.6 and 7.7 respectively. Solving equations 9.5 for $h(1)$ and $h(2)$, and substituting the values for $p_y(i,j)$ and $p_{xy}(j)$, we have

$$h(1) = \frac{\sigma_x^2}{\frac{3}{2}\sigma_x^2 + \sigma_v^2} \tag{9.6}$$

$$h(2) = \frac{h(1)}{\sqrt{2}} \tag{9.7}$$

Using these solutions in equation 7.1 we obtain the estimate of x as

$$\hat{x} = \frac{\sigma_x^2}{\frac{3}{2}\sigma_x^2 + \sigma_v^2}\,y(1) + \frac{1}{\sqrt{2}}\,\frac{\sigma_x^2}{\frac{3}{2}\sigma_x^2 + \sigma_v^2}\,y(2) \tag{9.8}$$

where $y(1)$ and $y(2)$ are the measurement samples.

9.2 Scalar Kalman filter

Estimate the value of a constant x, given discrete measurements of x corrupted by an uncorrelated Gaussian noise sequence with zero mean and variance σ_v^2.

The scalar equations describing this situation are

$$x(k) = x(k-1)$$

for the system and

$$y(k) = x(k) + v(k)$$

for the measurement, where

$$E[v(k)] = 0, \qquad E[v^2(k)] = \sigma_v^2$$

In this problem $a = 1$, and $w(k-1) = 0$, therefore $p_1(k) = p(k-1)$, see equation 7.51. Also, $c = 1$, so that the filter (or Kalman) gain, from equation 7.50, is given by

$$b(k) = \frac{p(k-1)}{p(k-1) + \sigma_v^2} \tag{9.9}$$

The error covariance, given by equation 7.52, in this case becomes

$$p(k) = \frac{p(k-1)}{1 + p(k-1)/\sigma_v^2}$$

which is a difference equation. This equation can be solved by noting that

$$p(1) = \frac{p(0)}{1 + p(0)/\sigma_v^2}$$

$$p(2) = \frac{p(1)}{1 + p(1)/\sigma_v^2} = \frac{p(0)}{1 + 2p(0)/\sigma_v^2}$$

$$\vdots$$

$$p(n) = \frac{p(0)}{1 + np(0)/\sigma_v^2} \tag{9.10}$$

Having obtained the above solution, we use it to calculate

$$b(k) = \frac{p(0)}{\sigma_v^2 + kp(0)} \tag{9.11}$$

Now we can write for the optimum (Kalman) discrete filter, described by equation 7.49, for this example:

$$\hat{x}(k) = \hat{x}(k-1) + \frac{p(k-1)}{\sigma_v^2 + p(k-1)}[y(k) - \hat{x}(k-1)] \tag{9.12}$$

where we have used equation 9.9 for b. From the above result we see that without measurement noise, i.e. $\sigma_v^2 = 0$, we have $\hat{x}(k) = y(k)$. But with noise, it is better to use equation 9.11 for b, in which case the Kalman filter is given by

$$\hat{x}(k) = \hat{x}(k-1) + \frac{p(0)}{\sigma_v^2 + kp(0)}[y(k) - \hat{x}(k-1)] \tag{9.13}$$

For sufficiently large k, we have $\hat{x}(k) = \hat{x}(k-1) = \hat{x}$, indicating that further measurements provide no new information.

9.3 Vector Kalman filter

Consider the second-order message model given by

$$\mathbf{x}(k) = \begin{bmatrix} 1 & 1 \\ 0 & 1 \end{bmatrix} \mathbf{x}(k-1) + \mathbf{w}(k-1)$$

The state is observed by means of a scalar observation model

$$y(k) = x_1(k) + v(k) \tag{9.14}$$

The input noise is stationary with

$$Q(k) = Q = \begin{bmatrix} 0 & 0 \\ 0 & 1 \end{bmatrix}$$

and the measurement noise is non-stationary, with $R(k) = 2 + (-1)^k$. In other words, the measurements for even values of k have more noise than the odd values of k. We assume that the variance of the initial errors (or initial state) is given by

$$P(0) = \begin{bmatrix} 10 & 0 \\ 0 & 10 \end{bmatrix}$$

We wish to compute the values of $K(k)$ for $k = 1$ to 10.

We start with the covariance equation 8.24, i.e.

$$P_1(k) = A\,P(k-1)A^{\mathrm{T}} + Q(k-1)$$

which for this problem becomes

$$P_1(k) = \begin{bmatrix} 1 & 1 \\ 0 & 1 \end{bmatrix} P(k-1) \begin{bmatrix} 1 & 0 \\ 1 & 1 \end{bmatrix} + \begin{bmatrix} 0 & 0 \\ 0 & 1 \end{bmatrix}$$

For $k = 1$, we have

$$P_1(1) = \begin{bmatrix} 1 & 1 \\ 0 & 1 \end{bmatrix} \begin{bmatrix} 10 & 0 \\ 0 & 10 \end{bmatrix} \begin{bmatrix} 1 & 0 \\ 1 & 1 \end{bmatrix} + \begin{bmatrix} 0 & 0 \\ 0 & 1 \end{bmatrix}$$

where we have substituted given initial error covariance $P(0)$. The above results in

$$P_1(1) = \begin{bmatrix} 20 & 10 \\ 10 & 11 \end{bmatrix}$$

which is used in equation 8.23 as follows

$$K(1) = P_1(1)C^{\mathrm{T}}[CP_1(1)C^{\mathrm{T}} + R(1)]^{-1}$$

$$\text{or} \quad K(1) = \begin{bmatrix} 20 & 10 \\ 10 & 11 \end{bmatrix} \begin{bmatrix} 1 \\ 0 \end{bmatrix} \left\{ [1 \quad 0] \begin{bmatrix} 20 & 10 \\ 10 & 11 \end{bmatrix} \begin{bmatrix} 1 \\ 0 \end{bmatrix} + \begin{bmatrix} 1 \\ 0 \end{bmatrix} \right\}^{-1}$$

$$= \begin{bmatrix} 20 \\ 10 \end{bmatrix} (21)^{-1} = \begin{bmatrix} 20/21 \\ 10/21 \end{bmatrix} = \begin{bmatrix} 0.95 \\ 0.48 \end{bmatrix}$$

giving $K_1(1) = 0.95$ and $K_2(1) = 0.48$.

Note that in the above calculations we have used

$$C = [1 \quad 0]$$

from the observation equation 9.14, from which we also have for the noise variance

$$R(k) = E[v(k)v^{\mathrm{T}}(k)]$$

$$= E\left\{ \begin{bmatrix} v_1(k) \\ 0 \end{bmatrix} [v_1{}^{\mathrm{T}}(k) \quad 0] \right\}$$

$$= \begin{bmatrix} E[v_1(k)v_1{}^{\mathrm{T}}(k)] \\ 0 \end{bmatrix} = \begin{bmatrix} R(k) \\ 0 \end{bmatrix}$$

where $R(k) = 2 + (-1)^k$.

To compute $K(2)$, we have first to find the value $P(1)$ from equation 8.25 in the following way:

$$P(k) = [1 - K(k)C]\,P_1(k)$$

$$P(1) = \left\{ \begin{bmatrix} 1 & 0 \\ 0 & 1 \end{bmatrix} - \begin{bmatrix} 20/21 \\ 10/21 \end{bmatrix} [1 \quad 0] \right\} \begin{bmatrix} 20 & 10 \\ 10 & 11 \end{bmatrix}$$

and after evaluations, we have

$$P(1) = \begin{bmatrix} 0.95 & 0.48 \\ 0.48 & 6.24 \end{bmatrix}$$

Now we use the above value of $P(1)$ to calculate the next value for $P_1(2)$ from

$$P_1(2) = AP(1)A^T + Q$$

$$= \begin{bmatrix} 1 & 1 \\ 0 & 1 \end{bmatrix} \begin{bmatrix} 0.95 & 0.48 \\ 0.48 & 6.24 \end{bmatrix} \begin{bmatrix} 1 & 0 \\ 1 & 1 \end{bmatrix} + \begin{bmatrix} 0 & 0 \\ 0 & 1 \end{bmatrix}$$

$$P_1(2) = \begin{bmatrix} 8.1 & 6.7 \\ 6.6 & 7.2 \end{bmatrix}$$

This result enables us to calculate the set of values for $K(2)$ from

$$K(2) = P_1(2)C^T[CP_1(2)C^T + R(2)]^{-1}$$

$$= \begin{bmatrix} 8.1 & 6.7 \\ 6.6 & 7.2 \end{bmatrix} \begin{bmatrix} 1 \\ 0 \end{bmatrix} \left\{ [1 \quad 0] \begin{bmatrix} 8.1 & 6.7 \\ 6.6 & 7.2 \end{bmatrix} \begin{bmatrix} 1 \\ 0 \end{bmatrix} + \begin{bmatrix} 3 \\ 0 \end{bmatrix} \right\}^{-1}$$

$$= \begin{bmatrix} 0.73 \\ 0.6 \end{bmatrix}$$

giving $K_1(2) = 0.73$ and $K_2(2) = 0.6$.

Next we compute $P(2)$ to obtain $P_1(3)$ and then $K(3)$, and so on. The components of $K(k)$ are shown in fig. 9.1 for $k = 1$ to 10. Note how the gain increases for odd values of k, to reflect the less noisy measurements. We see that the gain has reached an approximately periodic steady-state solution after only a few samples.

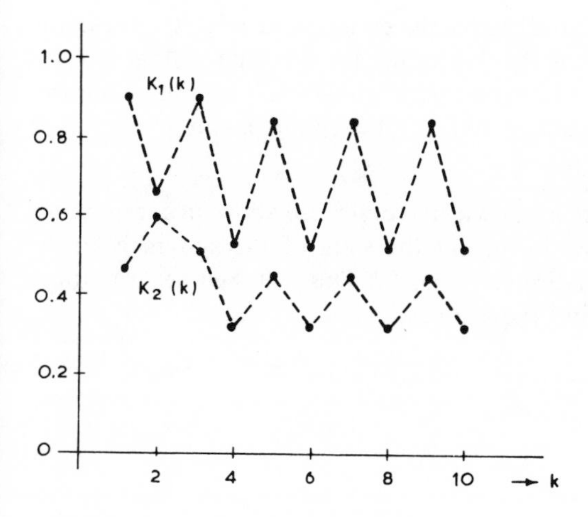

Fig. 9.1 Gain of vector Kalman filter

9.4 Kalman filter application to falling body

Consider a noise free second-order system representing a falling body in a constant field

$$\ddot{z} = -g \qquad t \geq 0 \tag{9.15}$$

Let the position be $z = x_1$, and velocity $\dot{z} = x_2$.

Then, defining the vector

$$\mathbf{x}(t) = \begin{bmatrix} x_1(t) \\ x_2(t) \end{bmatrix}$$

equation 9.15 can be written in state-space form

$$\dot{\mathbf{x}}(t) = \begin{bmatrix} 0 & 1 \\ 0 & 0 \end{bmatrix} x(t) + \begin{bmatrix} 0 \\ -g \end{bmatrix} \tag{9.16}$$

The state transition matrix of this system is

$$\mathbf{\Phi}(t, \tau) = \begin{bmatrix} 1 & t-\tau \\ 0 & 1 \end{bmatrix} \tag{9.17}$$

so that the solution for $\mathbf{x}(t)$ is given by

$$\mathbf{x}(t) = \mathbf{\Phi}(t, \tau)\mathbf{x}(\tau) + \int_\tau^t \begin{bmatrix} 0 & t-s \\ 0 & 1 \end{bmatrix} \begin{bmatrix} 0 \\ -g \end{bmatrix} \mathrm{d}s$$

or
$$\mathbf{x}(t) = \mathbf{\Phi}(t, \tau)\mathbf{x}(\tau) - g \begin{bmatrix} (t-\tau)^2/2 \\ (t-\tau) \end{bmatrix} \tag{9.18}$$

Substituting for $\mathbf{\Phi}(t, \tau)$ we can write the solution as

$$x_1(t) = x_1(\tau) + (t-\tau)x_2(\tau) - \frac{g}{2}(t-\tau)^2$$
$$x_2(t) = x_2(\tau) - g(t-\tau)$$

where the first equation is the body position, and the second equation its velocity. These are exact solutions which we have used to calculate the true values for $x_1(t)$ and $x_2(t)$ at times $t = 1, 2, \ldots, 6$. The results are shown in the first two columns in Table 9.1. It has been assumed that the 'true' initial conditions are $x_1(0) = z(0) = 100$, $x_2(0) = \dot{z}(0) = 0$, and $g = 1$.

Application of Kalman filter

The general dynamic behaviour of the system is known, and described above in continuous-time. We can easily obtain its discrete-time form, but the state of the system is to be determined from a set of measurements as indicated in fig. 9.2. This will then be a situation where we can use vector Kalman filter to find the state estimates.

We first convert system equations to discrete-time form by setting $t = kT$, $\tau = (k-1)T$, and taking $T = 1$. Applying this to equations 9.18, 9.17, we obtain

$$\mathbf{x}(k) = \underbrace{\begin{bmatrix} 1 & 1 \\ 0 & 1 \end{bmatrix}}_{A} \mathbf{x}(k-1) + \underbrace{\begin{bmatrix} \frac{1}{2} \\ 1 \end{bmatrix}}_{B} \underbrace{(-g)}_{\mathbf{u}} \tag{9.19}$$

which is the difference equation representing our system, i.e. a falling body in a constant field.

Table 9.1 Falling body in a constant field. The initial state values $\hat{x}(0)$ are given as the estimates at $t = kT = 0$, at which time the assumed errors in estimates are $p_{11}(0)$ and $p_{22}(0)$

	True values			*Estimates*		*Errors in Estimates*	
	Position	*Velocity*	*Position Observations*	*Position*	*Velocity*	*Position*	*Velocity*
$t = kT$	$x_1(t)$	$x_2(t)$	$y(k)$	$\hat{x}_1(k)$	$\hat{x}_2(k)$	$p_{11}(k)$	$p_{22}(k)$
0	100·0	0		95·0	1·0	10·0	1·0
1	99·5	−1·0	100·0	99·63	0·38	0·92	0·92
2	98·0	−2·0	97·9	98·43	−1·16	0·67	0·58
3	95·5	−3·0	94·4	95·21	−2·91	0·66	0·30
4	92·0	−4·0	92·7	92·35	−3·70	0·61	0·15
5	87·5	−5·0	87·3	87·68	−4·84	0·55	0·08
6	82·0	−6·0	82·1	82·22	−5·88	0·50	0·05

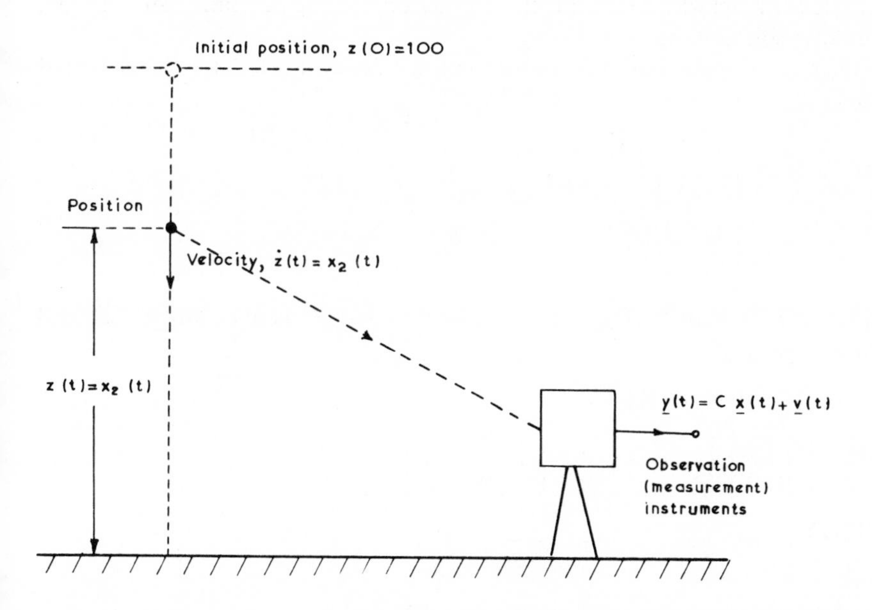

Fig. 9.2 Measurements of falling body for application of Kalman filter

This system has been observed by measuring the falling body position as in fig. 9.2, and the results of measurements are shown in the third column of Table 9.1. They have been affected by some independent random disturbance $v(k)$, so that the observation equation, in this case, can be written as

$$y(k) = \underbrace{[1 \quad 0]}_{C}\,\mathbf{x}(k) + v(k) \tag{9.20}$$

with the random disturbance (or noise) variance given by $R(k) = \sigma_v^2 = 1$.

Furthermore, we assume that the initial value of the state vector (position and velocity) is

$$\hat{\mathbf{x}}(0) = \begin{bmatrix} 95 \\ 1 \end{bmatrix}$$

and $\quad P(0) = \begin{bmatrix} 10 & 0 \\ 0 & 1 \end{bmatrix}$

are the assumed errors at time $k = 0$. Matrices A and C are indicated in equations 9.19 and 9.20. It will be noticed that we have, in this case, another term denoted by $B\mathbf{u}$. This is the driving force term which has been left out in our system equation 8.15, but we can easily incorporate it in the Kalman algorithm because it changes only the predictor term in equations 8.22 and 8.28, see also fig. 8.3, in the following way:

$$\hat{\mathbf{x}}'(k) = A\hat{\mathbf{x}}(k-1) + B\mathbf{u} \tag{9.21}$$

Therefore, the Kalman filter equation (8.22), for this case, is written as

$$\hat{\mathbf{x}}(k) = \hat{\mathbf{x}}'(k) + K(k)[y(k) - C\hat{\mathbf{x}}'(k)] \tag{9.22}$$

where $\hat{\mathbf{x}}'(k)$ is given by equation 9.21. All other equations for $P_1(k)$, $K(k)$ and $P(k)$ are unaltered. (Analysis of model with deterministic and random forcing functions can be found, for example, in Sorenson (26), pp. 226–9.)

To calculate the state estimates we start with equation 8.24, with $Q = 0$ for this case, and proceed as follows

$$P_1(1) = AP(0)A^{\mathrm{T}} = \begin{bmatrix} 1 & 1 \\ 0 & 1 \end{bmatrix} \begin{bmatrix} 10 & 0 \\ 0 & 1 \end{bmatrix} \begin{bmatrix} 1 & 0 \\ 1 & 1 \end{bmatrix}$$
$$= \begin{bmatrix} 11 & 1 \\ 1 & 1 \end{bmatrix}$$

Then, using equation 8.23, with $P_1(1)$ given by the above result, C as in equation 9.20, and $R(k) = 1$, we have

$$K(1) = P_1(1)C^{\mathrm{T}}[CP_1(1)C^{\mathrm{T}} + R]^{-1}$$
$$= \begin{bmatrix} 11 & 1 \\ 1 & 1 \end{bmatrix} \begin{bmatrix} 1 \\ 0 \end{bmatrix} \left\{ [1 \quad 0] \begin{bmatrix} 11 & 1 \\ 1 & 1 \end{bmatrix} \begin{bmatrix} 1 \\ 0 \end{bmatrix} + 1 \right\}^{-1}$$
$$= \begin{bmatrix} 11/12 \\ 1/12 \end{bmatrix}$$

Next we calculate the prediction term in equation 9.22, i.e.

$$\hat{\mathbf{x}}'(1) = A\hat{\mathbf{x}}(0) + B\mathbf{u}$$
$$= \begin{bmatrix} 1 & 1 \\ 0 & 1 \end{bmatrix} \begin{bmatrix} 95 \\ 1 \end{bmatrix} + \begin{bmatrix} -0.5 \\ -1 \end{bmatrix}$$
$$= \begin{bmatrix} 95.5 \\ 0 \end{bmatrix}$$

where we have used $\hat{\mathbf{x}}(0)$ specified earlier and $B\mathbf{u}$ term from equation 9.19 with $g = 1$.

Now we have all the components required in equation 9.22, which gives us the state estimate (or filtering equation) as

$$\hat{\mathbf{x}}(1) = \hat{\mathbf{x}}'(1) + K(1)[y(1) - C\hat{\mathbf{x}}'(1)]$$
$$= \begin{bmatrix} 95.5 \\ 0 \end{bmatrix} + \begin{bmatrix} 11/12 \\ 1/12 \end{bmatrix} \left\{ 100 - [1 \quad 0] \begin{bmatrix} 95.5 \\ 0 \end{bmatrix} \right\}$$

$$\text{or} \quad \hat{\mathbf{x}}(1) = \begin{bmatrix} 99.6 \\ 0.37 \end{bmatrix}$$

therefore $\hat{x}_1(1) = 99.6$ and $\hat{x}_2(1) = 0.37$. This is the first set of values estimated from Kalman filtering and entered in Table 9.1. The errors of the above estimated pair of values are calculated from equation 8.25 as follows:

$$P(1) = P_1(1) - K(1)\,CP_1(1)$$

$$= \begin{bmatrix} 11 & 1 \\ 1 & 1 \end{bmatrix} - \begin{bmatrix} 11/12 \\ 1/12 \end{bmatrix} [1 \quad 0] \begin{bmatrix} 11 & 1 \\ 1 & 1 \end{bmatrix}$$

$$= \begin{bmatrix} 11/12 & 1/12 \\ 1/12 & 11/12 \end{bmatrix}$$

The diagonal terms $p_{11}(1) = p_{22}(1) = 11/12$ are the ones of interest to us, because they represent the mean-square errors of position and velocity estimates. These have been entered in Table 9.1 for $k = 1$. The first computational cycle has now been completed. The results obtained, $P(1)$ and $\mathbf{x}(1)$, are then used to compute $\mathbf{x}(2)$ and $P(2)$, and so on, until we reach $\mathbf{x}(6)$ and $P(6)$. The computer program given by Rousseliere (31) and in section 10 of the appendix with comments, has been used to calculate estimates and errors, and the results obtained are shown in Table 9.1.

In this problem we are observing the position, and as a result, the position error drops fast as soon as the first observation is processed. However, the velocity error does not decrease much until the second observation is processed, because two position observations are required to determine both components of the state vector. The system is such that velocity affects the position, but position does not affect the velocity. Therefore, the velocity must first be estimated accurately before good estimates of position can be obtained. The above, and also the effect of the initial error conditions on the filtering procedure, are discussed by Jazwinski (29) and by Rousseliere (31).

9.5 Wiener filter application to falling body

In a treatment similar to the scalar Wiener filter of examples 7.1 and 7.2, we start with the Wiener–Hopf equation. This means applying the multidimensional Wiener–Hopf equation 8.49 to the falling body described in section 9.4.

We start, therefore, by rewriting equation 8.49 as

$$\sum_{i=1}^{6} H(k, i)\, E[\mathbf{y}(i)\, \mathbf{y}^{\mathrm{T}}(j)] = E[\mathbf{x}(k)\, \mathbf{y}^{\mathrm{T}}(j)] \tag{9.23}$$

Since the system has two states, and we measure only one state, position, this equation reduces to

$$\sum_{i=1}^{6} H(k, i)\, E[y(i)\, y(j)] = E\begin{bmatrix} x_1(k)\, y(j) \\ x_2(k)\, y(j) \end{bmatrix} \tag{9.24}$$

Introducing $p_y(i, j) = E[y(i)\, y(j)]$, $p_{x_1y}(k, j) = E[x_1(k)\, y(j)]$, $p_{x_2y}(k, j) = E[x_2(k)\, y(j)]$, and splitting $H(k, i)$ into two parts to correspond to the right-hand-side, we have

$$\sum_{i=1}^{6} \begin{bmatrix} H_1(k, i) \\ H_2(k, i) \end{bmatrix} p_y(i, j) = \begin{bmatrix} p_{x_1y}(k, j) \\ p_{x_2y}(k, j) \end{bmatrix} \tag{9.25}$$

From the above we obtain two equations

$$\sum_{i=1}^{6} H_1(k, i)\, p_y(i, j) = p_{x_1y}(k, j) \tag{9.26}$$

$$\sum_{i=1}^{6} H_2(k, i)\, p_y(i, j) = p_{x_2y}(k, j) \tag{9.27}$$

These equations must be solved for $H_1(k, i)$ and $H_2(k, i)$, which are required for calculations of the state estimates ($\hat{x}_1$ and $\hat{x}_2$) and their errors. The estimator equation 8.54 is split into two parts, producing the following two equations

$$\hat{x}_1(k) = \sum_{i=1}^{6} H_1(k, i)\, y(i) \tag{9.28}$$

$$\hat{x}_2(k) = \sum_{i=1}^{6} H_2(k, i)\, y(i) \tag{9.29}$$

The mean-square errors are given by the diagonal terms in the error covariance matrix, equation 8.55, which in this case are

$$p_{11}(k) = E[x_1^2(k)] - \sum_{i=1}^{6} H_1(k, i)\, E[y(i)\, x_1(k)] \tag{9.30}$$

$$p_{22}(k) = E[x_2^2(k)] - \sum_{i=1}^{6} H_2(k, i)\, E[y(i)\, x_2(k)] \tag{9.31}$$

As before, we introduce the notation $p_{x_1}(k, k) = E[x_1^2(k)]$, $p_{x_2}(k, k) = E[x_2^2(k)]$, $p_{yx_1}(i, k) = E[y(i)\, x_1(k)]$, and $p_{yx_2}(i, k) = E[y(i)\, x_2(k)]$, so the above equations become

$$p_{11}(k) = p_{x_1}(k, k) - \sum_{i=1}^{6} H_1(k, i)\, p_{yx_1}(i, k) \tag{9.32}$$

$$p_{22}(k) = p_{x_2}(k, k) - \sum_{i=1}^{6} H_2(k, i)\, p_{yx_2}(i, k) \tag{9.33}$$

We proceed now by using the measurement equation $y(i) = x_1(i) + v(i)$ to calculate the correlation quantities in equations 9.26 and 9.27. These are given as

$$\begin{aligned} p_{x_1y}(k, j) &= E[x_1(k)\, y(j)] = E\{x_1(k)[x_1(j) + v(j)]\} \\ &= E[x_1(k)\, x_1(j)] = p_{x_1}(k, j) \end{aligned} \tag{9.34}$$

$$\begin{aligned} p_{x_2y}(k, j) &= E\{x_2(k)\, [x_1(j) + v(j)]\} \\ &= E[x_2(k)\, x_1(j)] = p_{x_2x_1}(k, j) \end{aligned} \tag{9.35}$$

$$\begin{aligned} p_y(i, j) &= E\{[x_1(i) + v(i)]\, [x_1(j) + v(j)]\} \\ &= E[x_1(i)\, x_1(j)] + E[v(i)\, v(j)] \\ &= p_{x_1}(i, j) + \sigma_v^2\, \delta(i, j) \end{aligned} \tag{9.36}$$

where $\delta(i, j) \begin{cases} = 1 & i = j \\ = 0 & i \neq j \end{cases}$ (9.37) and $\sigma_v^2 = 1$ as specified in the previous section. In the error terms, equations 9.32 and 9.33, $p_{x_1}(k, k)$ is a special case of $p_{x_1}(k, j)$ for $k = j$, while $p_{x_2}(k, k)$ will be calculated separately. Furthermore, we have

$$p_{yx_1}(i, k) = E\{[x_1(i) + v(i)]\, x_1(k)\} = E[x_1(i)\, x_1(k)] = p_{x_1}(i, k)$$
$$p_{yx_2}(i, k) = E\{[x_1(i) + v(i)]\, x_2(k)\} = E[x_1(i)\, x_2(k)] = p_{x_1x_2}(i, k)$$

These two quantities can be expressed in terms of the previous quantities, i.e. $p_{x_1}(i, k) = p_{x_1}(k, i)$ and also $p_{x_1x_2}(i, k) = p_{x_2x_1}(k, i)$. The last equation means, for example, $E[x_1(i)\, x_2(k)] = E[x_2(k)\, x_1(i)]$ which is correct since the quantities within the square brackets are scalars.

From equations 9.34 to 9.36, we see that all of the required quantities are expressed in terms of p_{x_1} and $p_{x_2x_1}(k, j)$. To calculate these correlation terms, we use the system model, equation 9.19, which is rewritten for $g = 1$ as

$$\mathbf{x}(k) = \begin{bmatrix} 1 & 1 \\ 0 & 1 \end{bmatrix} \mathbf{x}(k-1) - \begin{bmatrix} \frac{1}{2} \\ 1 \end{bmatrix} \tag{9.38}$$

This is a particular case of the general first order vector equation

$$\mathbf{x}(k) = A\, \mathbf{x}(k-1) + B\mathbf{u}(k)$$

whose solution is

$$\mathbf{x}(k) = A^k\, \mathbf{x}(0) + \sum_{i=0}^{k-1} A^i\, B\mathbf{u}(k-i-1) \tag{9.39}$$

see, for example, Freeman (6), p. 19. Therefore, we can write the solution of equation 9.38 as follows

$$\mathbf{x}(k) = \begin{bmatrix} 1 & 1 \\ 0 & 1 \end{bmatrix}^k \mathbf{x}(0) - \sum_{i=0}^{k-1} \begin{bmatrix} 1 & 1 \\ 0 & 1 \end{bmatrix}^i \begin{bmatrix} \frac{1}{2} \\ 1 \end{bmatrix}$$

$$\text{or} \quad \mathbf{x}(k) = \begin{bmatrix} 1 & k \\ 0 & 1 \end{bmatrix} \mathbf{x}(0) - \begin{bmatrix} k^2/2 \\ k \end{bmatrix} \tag{9.40}$$

where the first term is obtained by application of the Cayley–Hamilton theorem (matrix algebra), and similarly the second term where a summation of the arithmetic progression has also been performed. The individual state vector components, from the above result, are given by

$$x_1(k) = x_1(0) + k\, x_2(0) - k^2/2 \tag{9.41}$$
$$x_2(k) = x_2(0) - k \tag{9.42}$$

where $x_1(0)$ and $x_2(0)$ are the initial position and velocity respectively. These values are unknown, and will be considered as random variables with mean values for the position $E[x_1(0)] = p_0$, velocity $E[x_2(0)] = v_0$, and variances p_1 and p_2 respectively. We also assume they are not correlated, i.e. $E[x_1(0)\, x_2(0)] = p_0 v_0$.

Now we can calculate $p_{x_1}(i, k)$ and $p_{x_2x_1}(k, j)$ as follows:

$$\begin{aligned} p_{x_1}(k, j) &= E\{[x_1(0) + k\, x_2(0) - k^2/2]\, [x_1(0) + j\, x_2(0) - j^2/2]\} \\ &= (p_0 + kv_0 - k^2/2)\, (p_0 + jv_0 - j^2/2) + p_1 + kjp_2 \end{aligned} \tag{9.43}$$

$$\begin{aligned} p_{x_2x_1}(k, j) &= E\{[x_2(0) - k]\, [x_1(0) + j\, x_2(0) - j^2/2]\} \\ &= (v_0 - k)\, (p_0 + jv_0 - j^2/2) + jp_2 \end{aligned} \tag{9.44}$$

since $E[x_i^2(0)] = [Ex_i(0)]^2 + p_i$ (9.45)

The first terms in equations 9.32 and 9.33 are obtained from equation 9.43, by setting $j = k$ and using $p_{x_2}(k, k)$ from

$$p_{x_2}(k, k) = E[x_2^2(k)] = (v_0 - k)^2 + p_2 \tag{9.46}$$

where equation 9.42 has also been used.

A computer program, obtained from Rousseliere (31) and used to calculate the position and velocity estimates and their mean-square errors, is based on the above preparatory work and shown in section 11 of the appendix with comments. Results for the initial conditions, $x_0 = 95$, $v_0 = 0$, and their variances $p_{11}(0) = p_1 = 10$, $p_{22}(0) = p_2 = 1$, are shown in Table 9.2.

Table 9.2 Velocity estimates and mean-square errors for Wiener filter applied to falling body

k	$\hat{x}_1(k)$	$\hat{x}_2(k)$	$p_{11}(k)$	$p_{22}(k)$
0	95	0·00	10	1
1	99·31	0·07	0·49	0·99
2	99·45	−0·91	45·94	1·15
3	91·64	−1·89	11·66	1·31
4	89·89	−2·88	7·80	1·49
5	86·43	−3·86	2·89	1·68
6	82·09	−4·84	1·00	1·88

Comparing these results with the Kalman filter results in Table 9.1, we see that the Kalman filter is a better estimator. A more complete comparison of Wiener, Kalman and the α–β method (given by equations 1.8), for initial conditions of position (100, 105) and velocity (−1, 0), has been made by Rousseliere (31).

The derivations and discussion of Wiener and Kalman filters in chapter 7 give the impression that the Wiener filter does not require a system equation, while the Kalman filter requires one which is of the form of a first-order recursive difference equation. This is also true because the Wiener filter has been derived for a constant signal x, while the Kalman filter has been developed for a time-varying random signal $x(k)$. However, in order to apply the Wiener filter to simple examples 7.1 and 7.2 a number of assumptions were made in order to calculate p_y and p_{xy}, and in example 7.2 the system was in fact described as a straight line (kx).

In this section, the application of the vector Wiener filter to the problem of a falling body, which is a dynamic system, shows quite strongly that to solve the problem we must use the system equation. This means that the Wiener filter requires, like the Kalman filter, a knowledge of the system dynamics, unless the autocorrelation and cross-correlation matrices (P_y and P_{xy}) can be obtained by some other means.

9.6 Kalman filter formulation for radar tracking

This example shows what must be done prior to the application of the Kalman filter algorithm to radar tracking. A brief introduction is given in the paragraph below.

In radar tracking, the time delay between transmission and reception of the pulses provides an estimate of the aircraft range (radial distances), while the location of the antenna beam at the time of detection provides the aircraft bearing (azimuth). A short range radar rotates typically at a scan rate of 15 revolutions/minute (r.p.m.), while longer range radars rotate at 6 r.p.m. Therefore, we have for these two cases new range and bearing estimates every 4 s and 10 s respectively. This means that the tracking filters are updated at such an interval, corresponding to the time interval previously denoted by T.

We have already considered a radar tracking case in chapter 8, examples 8.3 and 8.5. In example 8.3 we derived the system model, for variations about average values, with states $x_1(k) = \rho(k)$ for the range and $x_2(k) = \dot{\rho}(k)$ for the radial velocity. We now add two more states concerned with the bearing $x_3(k) = \theta(k)$, and bearing rate (or angular velocity) $x_4(k) = \dot{\theta}(k)$. The addition of these two states augments equation 8.11 into the following system equation:

$$\underbrace{\begin{bmatrix} x_1(k+1) \\ x_2(k+1) \\ x_3(k+1) \\ x_4(k+1) \end{bmatrix}}_{\mathbf{x}(k+1)} = \underbrace{\begin{bmatrix} 1 & T & 0 & 0 \\ 0 & 1 & 0 & 0 \\ 0 & 0 & 1 & T \\ 0 & 0 & 0 & 1 \end{bmatrix}}_{A} \underbrace{\begin{bmatrix} x_1(k) \\ x_2(k) \\ x_3(k) \\ x_4(k) \end{bmatrix}}_{\mathbf{x}(k)} + \underbrace{\begin{bmatrix} 0 \\ u_1(k) \\ 0 \\ u_2(k) \end{bmatrix}}_{\mathbf{w}(k)} \tag{9.47}$$

In this expression the noise terms $u_1(k)$ and $u_2(k)$ represent the change in radial velocity and bearing rate respectively over interval T. They are each T times radial and angular acceleration. We assume $u_1(k)$ and $u_2(k)$ to be random with zero mean, and we also assume that they are uncorrelated both with each other and individually from one time interval to the next.

The radar sensors are assumed to provide noisy estimates of the range $\rho(k) = x_1(k)$, and bearing $\theta(k) = x_3(k)$ at time intervals T. At time k, the two sensor outputs are then

$$y_1(k) = x_1(k) + v_1(k)$$
$$y_2(k) = x_3(k) + v_2(k)$$

Therefore, the data vector, as discussed in chapter 8, equation 8.13 and example 8.5, can be written as

$$\underbrace{\begin{bmatrix} y_1(k) \\ y_2(k) \end{bmatrix}}_{\mathbf{y}(k)} = \underbrace{\begin{bmatrix} 1 & 0 & 0 & 0 \\ 0 & 0 & 1 & 0 \end{bmatrix}}_{C} \underbrace{\begin{bmatrix} x_1(k) \\ x_2(k) \\ x_3(k) \\ x_4(k) \end{bmatrix}}_{\mathbf{x}(k)} + \underbrace{\begin{bmatrix} v_1(k) \\ v_2(k) \end{bmatrix}}_{\mathbf{v}(k)} \tag{9.48}$$

The additive noise, $\mathbf{v}(k)$, is usually assumed to be Gaussian with zero-mean and variances $\sigma_\rho^2(k)$ and $\sigma_\theta^2(k)$. So far, we have established vector equations for the system model given by equation 9.47, and data model given by equation 9.48. The next step is to formulate noise covariance matrices Q for the system, and R for the measurement model. For the latter, using equation 8.18, we have

$$R(k) = E[\mathbf{v}(k)\,\mathbf{v}^{\mathrm{T}}(k)] = \begin{bmatrix} \sigma_\rho^2(k) & 0 \\ 0 & \sigma_\theta^2(k) \end{bmatrix} \tag{9.49}$$

and the system noise covariance matrix, defined in equation 8.19, is for this case given by

$$Q(k) = E[\mathbf{w}(k)\,\mathbf{w}^{\mathrm{T}}(k)] = \begin{bmatrix} 0 & 0 & 0 & 0 \\ 0 & \sigma_1^2 & 0 & 0 \\ 0 & 0 & 0 & 0 \\ 0 & 0 & 0 & \sigma_2^2 \end{bmatrix} \tag{9.50}$$

where $\sigma_1^2 = E(u_1^2)$ and $\sigma_2^2 = E(u_2^2)$ are the variances of T times the radial and angular acceleration respectively. Specific values must be substituted for those variances in order to define the Kalman filter numerically. To simplify, we assume that the probability density function (p.d.f) of the acceleration in either direction (ρ or θ) is uniform and equal to $p(u) = 1/2M$, between limits $\pm M$; the variance is $\sigma_u^2 = M^2/3$. A more realistic p.d.f. is used by Schwartz & Shaw (21), based on Singer & Behnke (32). The variances in equation 9.50 are then $\sigma_1^2 = T^2\sigma_u^2$ and $\sigma_2^2 = \sigma_1^2/R^2$.

To start Kalman processing we have to initialize the gain matrix $K(k)$. For this purpose the error covariance matrix $P(k)$ has to be specified in some way. A reasonable *ad hoc* initialization can be established using two measurements, range and bearing, at times $k = 1$ and $k = 2$. From these four measurement data we can make the following estimates:

$$\hat{\mathbf{x}}(2) = \begin{bmatrix} \hat{x}_1(2) = \hat{\rho}(2) = y_1(2) \\ \hat{x}_2(2) = \hat{\dot{\rho}}(2) = \dfrac{1}{T}[y_1(2) - y_1(1)] \\ \hat{x}_3(2) = \hat{\theta}(2) = y_2(2) \\ \hat{x}_4(2) = \hat{\dot{\theta}}(2) = \dfrac{1}{T}[y_2(2) - y_2(1)] \end{bmatrix} \tag{9.51}$$

To calculate $P(2)$, we use the general expression for $P(k)$, equation 8.20, where for $k = 2$, we have

$$P(2) = E\{[\mathbf{x}(2) - \hat{\mathbf{x}}(2)][\mathbf{x}(2) - \hat{\mathbf{x}}(2)]^{\mathrm{T}}\} \tag{9.52}$$

Values for $\hat{\mathbf{x}}(2)$ are given by equation 9.51, and using equations 9.47 and 9.48 for $\mathbf{x}(2)$, we obtain the following result:

$$\mathbf{x}(2) - \hat{\mathbf{x}}(2) = \begin{bmatrix} & -v_1(2) \\ u_1(1) & -(v_1(2) - v_1(1))/T \\ & -v_2(2) \\ u_2(1) & -(v_2(2) - v_2(1))/T \end{bmatrix} \tag{9.53}$$

which is a (4×1) column vector.

In this case the error covariance matrix is a (4×4) matrix

$$P(2) = \begin{bmatrix} p_{11} & p_{12} & p_{13} & p_{14} \\ p_{21} & p_{22} & p_{23} & p_{24} \\ p_{31} & p_{32} & p_{33} & p_{34} \\ p_{41} & p_{42} & p_{43} & p_{44} \end{bmatrix} \tag{9.54}$$

Taking into account the independence of noise sources u and v, and also the independence between individual noise samples, it can be shown that the above matrix simplifies to

$$P(2) = \begin{bmatrix} p_{11} & p_{12} & 0 & 0 \\ p_{21} & p_{22} & 0 & 0 \\ 0 & 0 & p_{33} & p_{34} \\ 0 & 0 & p_{43} & p_{44} \end{bmatrix} \tag{9.55}$$

where
$$\begin{aligned} p_{11} &= \sigma_\rho^2 \\ p_{12} = p_{21} &= \sigma_\rho^2/T \\ p_{22} &= 2\sigma_\rho^2/T^2 + \sigma_1^2 \\ p_{33} &= \sigma_\theta^2 \\ p_{34} = p_{43} &= \sigma_\theta^2/T \\ p_{44} &= 2\sigma_\theta^2/T^2 + \sigma_2^2 \end{aligned} \tag{9.56}$$

As a numerical example, we take for range $R = 160$ km (= 100 miles), scan time $T = 15$ s, and a maximum acceleration $M = 2.1\ \text{ms}^{-2}$. Let the r.m.s. noise in the range sensor be equivalent to 1 km, therefore $\sigma_\rho = 10^3$ m. Furthermore, let r.m.s. noise σ_θ in the bearing sensor be 1° or 0.017 rad. These two figures define numerically the noise covariance matrix R.

For the above numerical values, we can calculate noise variances in the Q matrix, equation 9.50, as $\sigma_1^2 = 330$ and $\sigma_2^2 = 1.3 \times 10^{-8}$.

Using the above values in equation 9.55, we obtain the initial value of the estimation covariance matrix $P(2)$, or in the alternative notation $P(2|2)$,

$$P(2|2) = \begin{bmatrix} 10^6 & 6.7 \times 10^4 & 0 & 0 \\ 6.7 \times 10^4 & 0.9 \times 10^4 & 0 & 0 \\ 0 & 0 & 2.9 \times 10^{-4} & 1.9 \times 10^{-5} \\ 0 & 0 & 1.9 \times 10^{-5} & 2.6 \times 10^{-6} \end{bmatrix} \tag{9.57}$$

Since we have this error matrix at $k = 2$, we could try to use it to calculate the predictor gain $G(3)$ at $k = 3$, which is given by

$$G(3) = AP(3|2)C^{\text{T}}[CP(3|2)C^{\text{T}} + R]^{-1} \tag{9.58}$$

where all quantities (A, C, R) are known except $P(3|2)$. We might try to use equation 8.35 to calculate $P(3|2)$ as

$$P(3|2) = [A - G(2)C]P(2|1)A^{\text{T}} + Q$$

but $G(2)$ and $P(2|1)$ are not known. However, equation 8.35 has been derived from 8.24 and 8.25, so instead of equation 8.35, we can use 8.24 rewritten as

$$P(k|k-1) = AP(k-1|k-1)A^{\text{T}} + Q \tag{9.59}$$

which for $k = 3$ becomes

$$P(3|2) = AP(2|2)A^{\text{T}} + Q \tag{9.60}$$

where $P(2|2)$ is known from equation 9.57. Substituting for the other matrices from equations 9.47 and 9.50, and using the given numerical values we obtain

$$P(3|2) = \begin{bmatrix} 5\times 10^{6} & 2\times 10^{5} & 0 & 0 \\ 2\times 10^{5} & 9.3\times 10^{3} & 0 & 0 \\ 0 & 0 & 14.5\times 10^{-4} & 5.8\times 10^{-5} \\ 0 & 0 & 5.8\times 10^{-5} & 2.6\times 10^{-6} \end{bmatrix} \tag{9.61}$$

The diagonal values give us the prediction errors. The first and third elements are respectively mean-square range and bearing prediction errors for $k = 3$.

We are now able to calculate the predictor gain $G(3)$ using equation 9.61 in equation 9.58 which gives, after matrix manipulations, the following result:

$$G(3) = \begin{bmatrix} 1.33 & 0 \\ 3.3\times 10^{-2} & 0 \\ 0 & 1.33 \\ 0 & 3.3\times 10^{-2} \end{bmatrix} \tag{9.62}$$

The next step is to find $P(3|3)$ using equation 8.25 which for $k = 3$ gives

$$P(3|3) = P(3|2) - K(3)\,C\,P(3|2)$$

where $K(3) = A^{-1}\,G(3)$ (see section 8.4). The process is then repeated by finding $P(4|3)$, $G(4)$ etc. Graphical representation of complete results for this case can be found in Schwartz & Shaw (21), pp. 358–62.

Solutions to problems

Section 1.5

1.2 (a) $F(z) = 1 + \frac{1}{4}z^{-2} + \frac{1}{16}z^{-4}$; (b) $F(z) = \dfrac{1}{1 - \frac{1}{4}z^{-2}}$

1.4 $X(z) = z^2/(z-1)^2$; double zero at 0 and double pole at 1.

1.5 $Y(z)(1 + b_1 z^{-1}) = a_0 X(z)$; $Y(z)(1 - b_1 z^{-1} - b_2 z^{-2}) = a_0 X(z)$

1.7 All have poles at

$$z_{1,2} = (2 - \alpha - \beta)/2 \pm \tfrac{1}{2}(\beta^2 + \alpha^2 + 2\alpha\beta - 4\beta)^{1/2}$$

For critically damped system: $\alpha^2 + 2\alpha\beta + (\beta^2 - 4\beta) = 0$.

1.9 (b) $y = (6, 11, 4)$

Section 2.6

2.1 (a) $[y(k)] = [1, -1.5, 1.25, 1.625, -1.18]$; (b) $[y(k)] = [2, -1, \frac{3}{2}, \frac{7}{4}, -\frac{9}{8}]$

2.4 $y(k) = (-1)^k \dfrac{(1 - b_1^{k+1})}{[1 - b_1]} a_0$

2.5 $[x_1(k)] = [0, 1, 1, 1, 1]$; $[x_2(k)] = [1, \frac{5}{2}, \frac{5}{4}, \frac{5}{8}, \frac{5}{16}]$

$x_1(k) = 1 - \delta(k)$; $x_2(k) = 5(\frac{1}{2})^k - 4\delta(k)$.

2.6 (a) $[x(k)] = [1, 1, \frac{1}{2}, 0, -\frac{1}{4}, -\frac{1}{4}]$; (b) $x(k) = \dfrac{1}{(\sqrt{2})^k}\left(\cos\dfrac{k\pi}{4} + \sin\dfrac{k\pi}{4}\right)$

2.7 (a) $H(z) = \dfrac{0.5z^{-1}}{1 + 0.5z^{-1}}$; (b) $[h(k)] = [0, \frac{1}{2}, -\frac{1}{4}, \frac{1}{8}, -\frac{1}{16}, \frac{1}{32} \ldots]$

(c) $|H(\omega)| = \dfrac{0.45}{(1 + 0.8\cos\omega T)^{1/2}}$ (highpass); $\theta = -\omega T + \tan^{-1}\left(\dfrac{0.5\sin\omega T}{1 + 0.5\cos\omega T}\right)$

2.8 $|H_1(\omega)| = \left|\cos\dfrac{\omega T}{2}\right|$, $\theta_1 = -\omega T/2\,(z)$; $|H_2(\omega) = |\cos\omega T|$, $\theta_2 = -\omega T\,(z \to z^2)$

2.9 $|H_1(\omega)| = \sin(\omega T/2)$, $\theta_1 = (\pi/2) - (\omega T/2)$ (single unit);
$|H_2(\omega)| = \sin^2(\omega T/2)$, $\theta_2 = \pi - \omega T$ (double unit)

2.10 (a) $[y(k)] = [\frac{1}{4}, -\frac{1}{4}, 0, 0, 0, 0]$; (b) $[y(k)] = [0, \frac{1}{4}, 0, 0, 0, 0]$

2.11 (a) $y(k) = -\frac{1}{8}[\frac{1}{2}x(k) - x(k-1) + \frac{1}{2}x(k-2)] - \frac{5}{4}y(k-1) - \frac{1}{2}y(k-2)$

(b) $$|H(\omega)|^2 = \frac{1}{16}\frac{\sin^4(\omega T/2)}{(29/16) + (15/4)\cos\omega T + 2\cos^2\omega T}$$

2.13 $[h(k)] = [1, 2^{-1}, 0, 0, 0, 0, 2^{-6}, 2^{-7}, 0, 0, 0, 0, 2^{-12}, 2^{-13}, 0, 0, \ldots]$

2.14 (a) Double pole at $z_p = 1 - \sqrt{\beta}$. For stability, poles within the unit circle, $0 < \beta < 4$.
(b) $h(k) = (1-k)(0.1)^k, (\beta = 0.81)$; $h(k) = (0.2 - 3.6k)(-0.9)^k, (\beta = 3.6)$

Section 3.4

3.1 $[h(k)] = [0, -1, \frac{1}{2}, -\frac{1}{3}, \frac{1}{4}, -\frac{1}{5}, \frac{1}{6}, -\frac{1}{7}, \frac{1}{8}, -\frac{1}{9}]$
$h(-k) = h(k)$

3.2 $|H(\omega)|_3 = 2\sin\omega T; |H(\omega)|_5 = 2\sin\omega T - \sin 2\omega T$

3.3 (a) $$h(n) = \frac{1}{2}\left(\frac{\sin(n\pi/2)}{n\pi/2}\right)$$

(b) $$[h(k)]_R = \left[\frac{1}{2}, \frac{1}{\pi}, 0, -\frac{1}{3\pi}, 0, \frac{1}{5\pi}, 0\right]$$

$$[h(k)]_H = \left[\frac{1}{2}, \frac{(2-\sqrt{3})\alpha + \sqrt{3}}{2}\left(\frac{1}{\pi}\right), 0, \alpha\left(-\frac{1}{3\pi}\right), 0, \frac{(2\alpha + \sqrt{3})\alpha - \sqrt{3}}{2}\left(\frac{1}{5\pi}\right), 0\right]$$

3.4 (a) $|H(\omega)| = |0.25 + 0.46\cos\omega T|$; (b) $|H(\omega)| = |0.25 - 0.46\cos\omega T|$

3.5 (a) $|H(\omega)| = |0.50 - 0.64\cos 2\omega T|$; (b) $|H(\omega)| = |0.50 + 0.64\cos 2\omega T|$

Section 4.5

4.1 (b) $y(k) = (T/2\,e^{T/\sqrt{2}}\sin(T/\sqrt{2}))\,x(k-1) - (2\,e^{-T/\sqrt{2}}\cos(T/\sqrt{2}))\,y(k-1)$
$- e^{-T/\sqrt{2}}\,y(k-2)$

4.2 (b) $$y(k) = \frac{T^2}{D}x(k) + \frac{2T^2}{D}x(k-1) + \frac{T^2}{D}x(k-2) - \frac{2T^2-8}{D}y(k-1)$$
$$-\frac{4 - 2\sqrt{2}T + T^2}{D}y(k-2)$$

where $D = 4 + 2\sqrt{2}T + T^2$.
As $T \to 0$, the solutions for problems 4.1 and 4.2 reduce to

$$y(k) = T^2\,x(k-1) + 2\,y(k-1) - y(k-2)$$

where $x(k) = x(k-1), x(k-2) = x(k-1)$ has been used.

4.3 (a) $\dfrac{\omega_c}{\omega_c + s} \to \dfrac{\omega_c T}{1 - e^{-\omega_c T}z^{-1}}$; (b) All $\omega_c \to \omega_c T$.

4.4 (b) $$H(z) = \omega_c^2 T^2 \frac{1+2z^{-1}+z^{-2}}{[(4+2\sqrt{2}\,\omega_c T+\omega_c^2 T^2)+2(\omega_c^2 T^2-1)z^{-1} + (4-2\sqrt{2}\omega_c T+\omega_c^2 T^2)z^{-2}]}$$

4.5 (a) $$H(z) = \frac{1-2z^{-1}+z^{-2}}{(\beta^2+\beta\sqrt{2}+1)+(2\beta^2-2)z^{-1}+(\beta^2-\beta\sqrt{2}+1)z^{-2}}$$

$H(\text{highpass}) = H(\text{lowpass with } z^{-1} \to -z^{-1})$.

(b) $(2+\sqrt{2})\,y(k) = [x(k)+x(k-2)]-2x(k-1)-(2-\sqrt{2})\,y(k-2)$

4.6 $y(k) = \frac{1}{3}x(k)-\frac{2}{3}x(k-2)+\frac{1}{3}x(k-4)+\sqrt{2}\,y(k-1)-\frac{2}{3}y(k-2)+\frac{1}{3}\sqrt{2}y(k-3) -\frac{1}{3}y(k-4)$

4.7 $$(2+\sqrt{2})\,y(k) = [x(k)+x(k-4)]+4x(k-2)-2\sqrt{2}[x(k-1)+x(k-3)] + 2(1+\sqrt{2})\,y(k-1)-(2-\sqrt{2})[y(k-2)+y(k-4)] + 2(\sqrt{2}-1)\,y(k-3)$$

4.10 $\varepsilon_1 = \varepsilon_2 = 0.5\%$ for $\lambda_1 = 200\pi$; $\varepsilon_1 = \varepsilon_2 = 5\%$ for $\lambda_1 = 20\pi$
$\varepsilon_3 = 0.000832\%$ for $\lambda_1 = 200\pi$; $\varepsilon_3 = 0.083\%$ for $\lambda_1 = 20\pi$

Section 5.5

5.1 (a) $F = 4$ Hz; (b) $f_s = 2048$ Hz; (c) $f(\text{sig. max.}) = 1024$ Hz

5.2 $F_p(m\Omega) = e^{-jm\pi}2(1+\cos(m\pi/2))$

5.3 (a) $F_p(m) = e^{-j(2\pi/5)m}\dfrac{\sin(m\pi/2)}{\sin(m\pi/10)}$; (b) $F_p(m) = \dfrac{\sin(m\pi/2)}{\sin(m\pi/10)}$

5.4 (a) $F(m\Omega) = e^{-j(2\pi/3)m}\left(1-\cos\left(\dfrac{2\pi}{3}m\right)\right), \Omega = \dfrac{2\pi}{3}$

(b) $F(m\Omega) = e^{-j(2\pi/8)m}\left(1-\cos\left(\dfrac{2\pi}{8}m\right)\right), \Omega = \dfrac{2\pi}{8}$

5.5 (a) $g^{-1} = h = (\frac{1}{2}, -\frac{1}{4})$; $g^{-1} = h = (\frac{1}{2}, -\frac{1}{4}, \frac{1}{8})$
(b) Error energy: 1/16, for 2-length inverse.
1/64, for 3-length inverse.

5.6 $h_1 = g_1^{-1} = (\frac{1}{2}, -\frac{1}{4}, -\frac{1}{8})$, error energy: 5/32; $h_2 = g_2^{-1} = (1, -1, -1)$, error energy: 13.
The first sequence, the minimum delay type, has a stable inverse. However, for the second sequence, the maximum delay type, the inverse is unstable; see, for example, Robinson(8), p. 156.

Appendix

1 Derivation of second-order difference equation

Applying backward differences as shown by Hovanessian *et al.*(3) to equation 1.3, we obtain its discrete-time equivalent

$$\frac{y_n - 2y_{n-1} + y_{n-2}}{\Delta t^2} + 2\sigma \frac{y_n - y_{n-1}}{\Delta t} + \omega_0^2 y_n = \omega_0^2 x_n$$

or rewriting this, we have

$$y_n = \frac{2(1+\sigma\,\Delta t)}{D} y_{n-1} - \frac{1}{D} y_{n-2} + \frac{\omega_0^2\,\Delta t^2}{D} x_n$$

where $D = 1 + 2\sigma\Delta t + \omega_0^2\Delta t^2$.

The first two terms on the right-hand-side can be modified using the binomial expansion $(1+x)^{-1} \simeq 1 - x + x^2$. Substituting for $x = 2\sigma\,\Delta t + \omega_0^2\,\Delta t$ and squaring etc. we can 'force' (or 'massage') the coefficients of the above difference equation, written as

$$y_n = b_1\, y_{n-1} + b_2\, y_{n-2} + a_0\, x_n$$

into the following forms:

$$\begin{aligned} b_1 &= 2(1 - \sigma\,\Delta t + \ldots)[1 - (\omega_0^2 - \sigma^2)\,\Delta t^2 + \ldots] \\ &\simeq 2\,\mathrm{e}^{-\sigma\,\Delta t} \cos \Delta t\,\sqrt{(\omega_0^2 - \sigma^2)} \\ b_2 &= -(1 - 2\,\sigma\,\Delta t + \ldots) \simeq \mathrm{e}^{-2\sigma\,\Delta t} \end{aligned}$$

$$\text{and} \quad a_0 = \frac{1}{(1 + 2\,\sigma\,\Delta t)/\omega_0^2\,\Delta t^2 + 1} \simeq 1 \qquad \text{(A.1)}$$

since $\omega_0 \gg \sigma$, and assuming $(\omega_0\,\Delta t) \gg 1$.

2 Relation between *s*-plane and *z*-plane

A point $s_1 = \sigma_1 + \mathrm{j}\omega_1$ in the *s*-plane transforms to a point z_1 in the *z*-plane, defined by equation 1.19, so that we can write

$$z_1 = \mathrm{e}^{s_1 T} = \mathrm{e}^{\sigma_1 T} \mathrm{e}^{\mathrm{j}\omega_1 T}$$

This means $|z_1| = \mathrm{e}^{\sigma_1 T}$, and $\angle z_1 = \omega_1 T$, as shown in fig. A.1(a). Note that the location of point $\angle z_1$ changes as the sampling interval T is altered. To obtain more information on *s*–*z* plane relationship, we consider now the case $\sigma_1 = 0$, and find ω_1 for which $\angle z_1$ changes from $-\pi$ to $+\pi$, i.e. $-\pi < \angle z_1 < \pi$. Since $\angle z_1 = \omega_1 T$, we have further $-\pi < \omega_1 T < \pi$, or $-\pi/T < \omega_1 < \pi/T$, or $-(\omega_s/2) < \omega_1 < (\omega_s/2)$. Therefore, the path along the imaginary axis between $\pm(\omega_s/2)$ transforms into the unit circle in the *z*-plane, fig. A.1(b), since for $\sigma = 0$, $|z| = 1$. It follows that an l.h.s. strip ω_s wide in the *s*-plane transforms inside the unit circle ($\sigma < 0$). The corresponding r.h.s. strip ($\sigma > 0$) transforms outside the unit circle as

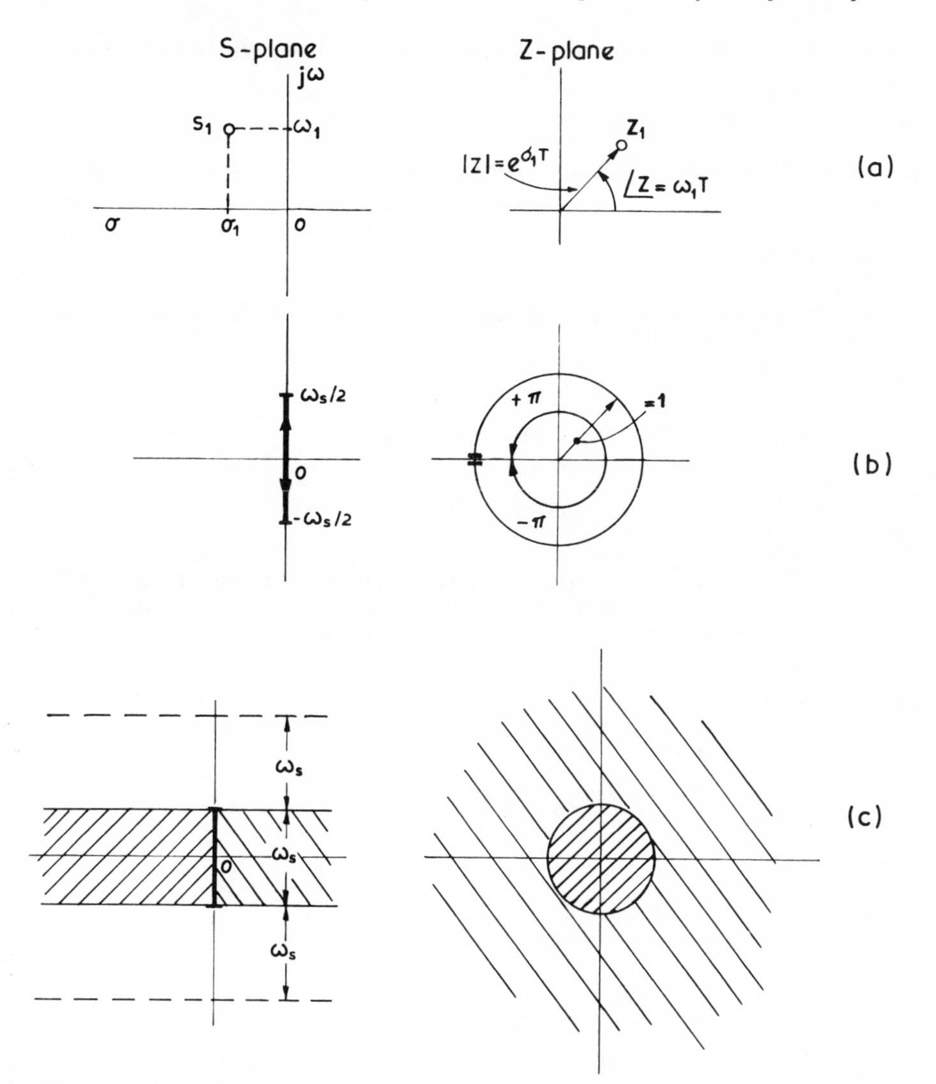

Fig. A.1 (a) Transformation of point in s-plane to point in z-plane; (b) transformation of imaginary axis in s-plane to unit circle in z-plane; (c) transformation of s-plane into z-plane

indicated in fig. A.1(c). Successive strips in the s-plane transform into the same place. Strictly, the strips transform into successive Riemann surfaces superimposed on the z-plane.

Note that the transformation used here, $z = e^{sT}$, is also the one used in control. In communications $z = e^{-sT}$ is used, and in this case the inside and outside regions in fig. A.1 (c) will have reversed roles.

3 Derivation of equation 1.20, the Laplace transform of a sampled wave

Assume that $F(s)$ is a rational polynomical in s with partial-fraction expansion given by

$$F(s) = \frac{A_1}{s+s_1} + \frac{A_2}{s+s_2} + \ldots = \sum_i \frac{A_i}{s+s_i} \tag{A.2}$$

The inverse Laplace transform gives us the corresponding time function as a sum of exponentials, i.e.

$$f(t) = \sum_i A_i e^{-s_i t}$$

To determine the sampled transform $F_s(s)$, we consider the first term of equation A.2 in the following way:

$$F_1(s) = \frac{A_1}{s+s_1} \to f_1(t) = A_1 e^{-s_1 t}$$

Using the sampled function $f_1(kT) = A_1 e^{-s_1 kT}$ in equation 1.17, we obtain the sampled transform

$$F_{s1}(s) = A_1 \sum_{k=0}^{\infty} e^{-s_1 kT} e^{-skT}$$

or summing this geometric series, we obtain the following closed form

$$F_{s1}(s) = \frac{A_1}{1 - e^{-(s+s_1)T}} \tag{A.3}$$

The poles of this function are given by $e^{-(s_p+s_1)T} = 1$, i.e. $-(s_{pn}+s_1)T = jn2\pi$, or $s_{pn} = -s_1 - jn(2\pi/T)$ with $n = 0, \pm 1, \pm 2, \ldots$. This shows that a pole at s_1, for a continuous time signal, will be repeated along a vertical line, for the sampled signal, at intervals $2\pi/T = \omega_s$, where ω_s is the sampling frequency in rad s^{-1}. We expand now $F_{s1}(s)$ into a sum of partial-fractions

$$F_{s1} = \sum_n \frac{c_n}{s - s_{pn}}$$

where c_n are evaluated from equation A.3 using

$$c_n = \frac{A_1}{\dfrac{d}{ds}[1 - e^{-(s+s_1)T}]}$$

at $s = s_{pj}$, resulting in $c_n = A_1/T$ at each pole.

Therefore, $F_{s1}(s)$ can be written in terms of its poles as

$$F_s(s) = \frac{A_1}{T}\left[\ldots \frac{1}{s+(s_1 - jn\omega_s)} + \ldots \frac{1}{s+(s_1 - j\omega_s)} + \frac{1}{[s+s_1]} + \frac{1}{s+(s_1+j\omega_s)} + \ldots \frac{1}{s+(s_1+jn\omega_s)} \ldots\right]$$

The same can be done for the other terms of equation A.2 with the result

$$\begin{aligned} F_s(s) &= \frac{A_1}{T}\left[\ldots \frac{1}{s+(s_1 - j\omega_s)} + \frac{1}{s+s_1} + \frac{1}{s+(s_1+j\omega_s)} + \ldots\right] \\ &+ \frac{A_2}{T}\left[\ldots \frac{1}{s+(s_2 - j\omega_s)} + \frac{1}{s+s_2} + \frac{1}{s+(s_2+j\omega_s)} + \ldots\right] \\ &+ \frac{A_i}{T}\left[\ldots \frac{1}{s+(s_i - j\omega_s)} + \frac{1}{s+s_i} + \frac{1}{s+(s_i+j\omega_s)} + \ldots\right] \\ &= \frac{1}{T}\left[\ldots F(s - j\omega_s) + F(s) + F(s + j\omega_s) + \ldots\right] \end{aligned}$$

or $$F_s(s) = \frac{1}{T} \sum_{n=-\infty}^{\infty} F(s + jn\omega_s)$$

which is given by equation 1.20. A more rigorous derivation of equations 1.16 and 1.20 can be found, for example, by Freeman(6), chapter 4.

4 Computer programs for nonrecursive filters

A FORTRAN program for calculation of Kaiser window coefficients given by

$$w_k(n) = \frac{I_0[2\pi\sqrt{(1-(n/10)^2)}]}{I_0(2\pi)}$$

using $I_0(x) = 1 + \sum_{k=1}^{K} \left[\frac{(x/2)^k}{k!}\right]^2$

```
0          PROGRAM    KCOEFF
1          END
2                     PI = 3.14159265
3                     T = 1. E-08
4                     DO 25 I = 1, 11
5                     G = I - 1
6                     Y = PI*SQRT(1. -0.01*G*G)
7                     E = 1.
8                     DE = 1.
9                     DO 10 J = 1, 100
10                    DE = DE*Y/FLOAT(J)
11                    SDE = DE*DE
12                    E = E + SDE
13                    IF (T - SDE) 10, 10, 20
14         10         CONTINUE
15                    J = 0.
16         20         IF (I .EQ. 1) E0 = E
17                    W = E/E0
18         25         WRITE (2,30) Y, E, W, J
19         30         FORMAT (10X, 3F15.10, I10)
20                    STOP
21                    END
```

A FORTRAN program for calculation of the nonrecursive filter response described by

$$10\log_{10}\frac{|H(m)|^2}{|H(0)|^2}$$

where $|H(m)|^2$ are given by expressions 3.17, 3.22 and 3.25, with results shown in figs 3.4 and 3.6.

```
0          PROGRAM WINDOWTEST
1          END
2                     DIMENSION A(25), SUM(5)
3                     READ (1,10) N
4          10         FORMAT (5X, I2)
5                     DO 70 J = 1, N
6                     READ (1,20) HINC, M, NA
7          20         FORMAT (5X, F5.2, 5X, I5, 5X, I1)
8                     WRITE (2,20) HINC, M, NA
9                     IF (NA .NE. 0) READ (1,25) (A(L), L = 1, M)
10                    WRITE (2,25) (A(L), L = 1, M)
11         25         FORMAT (1X, 5F15.10)
12                    PI = 3.14159265
13         30         H = 0.
14         35         DO 50 K = 1, 5
15                    IF (H .GT. 1) GO TO 70
16                    SUM(K) = 0.25
```

```
17               DO 4Ø I = 1, M
18               G = I
19         4Ø    SUM(K) = SUM(K) + A(I)*COS(G*H*PI)
2Ø               IF (H .LT. HINC) SUMØ = SUM(1)
21               SUMN = ABS(SUMØ/SUM(K))
22               SUM(K) = 2Ø.*ALOG1Ø(SUMN)
23               G = H
24         5Ø    H = H + HINC
25               WRITE (2,6Ø) G, SUM
26         6Ø    FORMAT (5X, F1Ø.4, 5X, 5F1Ø.3)
27               GO TO 35
28         7Ø    CONTINUE
29               STOP
3Ø               END
```

5 Computer program for recursive filters

A FORTRAN program for calculating

$$10\log_{10}\frac{|H(m)|^2}{|H(r)|^2} \text{ of figs 4.2, 4.4, 4.7, 4.8 and 4.9}$$

where $|H(m)|^2 = (\text{const})\dfrac{A_1 + A_2\cos m\pi + A_3\cos 2m\pi + A_4\cos 3m\pi}{A_5 + A_6\cos m\pi + A_7\cos 2m\pi + A_8\cos 3m\pi}$

$|H(r)|^2 = |H(0)|^2$ for lowpass filters (figs 4.2, 4.4, 4.7 and 4.8)
$|H(r)|^2 = |H(1)|^2$ for bandpass filter (fig. 4.9), and in this case m represents $2m$.

```
Ø     PROGRAM DFLTR
1     END
2                DIMENSION A(8)
3                PI = 3.14159265
4                READ (1,1Ø) N
5          1Ø    FORMAT (5X, I2)
6                DO 3Ø J = 1, N
7                READ (1,2Ø) C, A
8          2Ø    FORMAT (1X, F11.8/(4F15.1Ø)
9                WRITE (2,2Ø) C, A
1Ø               DO 3Ø M = 1Ø1, 5Ø1, 2
11               AM = M - 1
12               ANG = AM*PI/1ØØ.
13               TOP = A(1)+A(2)*COS(ANG)+A(3)*COS(2.*ANG)+A(4)*COS(3.*ANG)
14               DEN = A(5)+A(6)*COS(ANG)+A(7)*COS(2.*ANG)+A(8)*COS(3.*ANG)
15               HMSQ = C*TOP/DEN
16               IF (M .EQ. 1Ø1) HØSQ = HMSQ
17               HMSN = HMSQ/HØSQ
18               HMSN = ABS(HMSN)
19               IF (HMSN .LE. Ø) GO TO 5Ø
2Ø               RES = 1Ø.*ALOG1Ø(HMSN)
21               WRITE (2,4Ø) AM, RES
22         3Ø    CONTINUE
23               STOP
24         4Ø    FORMAT (5X, F5.1, 5X, F9.3)
25         5Ø    WRITE (2,6Ø) AM
26         6Ø    FORMAT (5X, F5.1, 5X, 11HFN INFINITE)
27               GO TO 3Ø
28               END
```

The above program is for fig. 4.9, and for figs 4.2, 4.4, 4.7 and 4.8 only step 10 should be changed to: DO 3Ø M1, 2Ø1, 5

6 Mean-square error of the sample mean estimator

Equation 6.16 is obtained from

$$p_e = E\left[\frac{1}{m}\sum_{i=1}^{m} v(i)\right]^2 = \frac{1}{m^2}\sum_{i=1}^{m}\sum_{j=1}^{m} E[v(i)v(j)]$$
$$= \frac{1}{m^2}\sum_{i=1}^{m}\sum_{j=1}^{m} \sigma_v^2 \delta_{ij}$$

where δ_{ij} represents the Kronecker delta, i.e. $\delta_{ij} = 1$ for $i = j$, and $\delta_{ij} = 0$ for $i \neq j$. Therefore, we have

$$\sum_{i=1}^{m}\sum_{j=1}^{m} \delta_{ij} = \sum_{i=1}^{m} (\delta_{i1} + \delta_{i2} + \ldots + \delta_{im})$$
$$= \delta_{11} + \delta_{12} + \ldots + \delta_{1m}$$
$$+ \delta_{21} + \delta_{22} + \ldots + \delta_{2m}$$
$$+ \delta_{m1} + \delta_{m2} + \ldots + \delta_{mm}$$

In the above only δ_{ij} for $i = j$ are equal to 1, all the others are zero. There are m such terms, therefore the double sum in this case reduces to m. The result is then $p_e = \sigma_v^2/m$.

7 Mean-square error for recursive filter

The error is

$$e = \hat{x} - x = -a^m x + (1-a)\sum_{i=1}^{m} a^{m-i} v(i)$$

and the mean-square error is

$$p_e = E(e^2) = E[-a^m x + (1-a)\sum_{i=1}^{m} a^{m-i} v(i)]^2$$
$$= a^{2m}E(x^2) + (1-a)^2 E[\sum_{i=1}^{m} a^{m-i} v(i)]^2$$

since $E[xv(i)] = 0$, because x and $v(i)$ are not correlated.

Further, we have

$$p_e = a^{2m}S + (1-a)^2 \sum_{i=1}^{m}\sum_{j=1}^{m} a^{m-i} a^{m-j} \sigma_v^2 \delta_{ij}$$

where we have represented $E[v(i)v(j)] = \sigma_v^2 \delta_{ij}$, as in section 6 of the appendix, and we denote $E(x^2)$ as S. Since $\delta_{ij} = 1$ only for $i = j$, the second term simplifies, so we can write

$$p_e = a^{2m}S + (1-a)^2 \sigma_v^2 \sum_{i=1}^{m} (a^{m-i})^2$$
$$= a^{2m}S + (1-a)^2 \sigma_v^2 [a^{2(m-1)} + a^{2(m-2)} + \ldots + a^2 + 1]$$

The square bracketed part is a geometric series, and summing this series we have

$$p_e = a^{2m}S + \frac{(1-a)(1-a^{2m})}{1+a}\sigma_v^2$$

8 Relationship between $a(k)$ and $b(k)$ in section 7.3.2

We start with $E[e(k)\hat{x}(k-1)] = 0$, and substitute $e(k) = \hat{x}(k) - x(k)$, and using equation 7.38 for $\hat{x}(k)$, we have

$$E\{[a(k)\hat{x}(k-1)+b(k)y(k)-x(k)]\hat{x}(k-1)\}=0$$

Adding and subtracting $a(k)x(k-1)$, the above becomes

$$E\{[a(k)[\hat{x}(k-1)-x(k-1)]+a(k)x(k-1)]\hat{x}(k-1)\} = E\{[x(k)-b(k)y(k)]\hat{x}(k-1)\}$$

This can be written as

$$a(k)E[e(k-1)\hat{x}(k-1)+x(k-1)\hat{x}(k-1)] = E\{[x(k)[1-cb(k)]-b(k)v(k)]\hat{x}(k-1)\}$$

where we have used equation 7.37 for $y(k)$.

The first term on the left-hand-side, $E[e(k-1)\hat{x}(k-1)]=0$, because we can write $\hat{x}(k-1) = a(k-1)\hat{x}(k-2)+b(k-1)y(k-1)$ and use the orthogonality equations 7.43 and 7.44 for the previous time $(k-1)$. The second term on the right-hand-side $E[v(k)\hat{x}(k-1)]=0$, because the estimate at time $(k-1)$ is not correlated with the observation noise at time k. Therefore, the above equation reduces to

$$a(k)E[x(k-1)\hat{x}(k-1)] = [1-cb(k)]E[x(k)\hat{x}(k-1)]$$

Substituting now for $x(k) = ax(k-1)+w(k-1)$ we have $E[w(k-1)\hat{x}(k-1)]=0$. To show this, we express

$$\hat{x}(k-1) = a(k-1)\hat{x}(k-2)+acb(k-1)x(k-2)+cb(k-1)w(k-2)+b(k-1)v(k-1)$$

where we have used equations 7.38, 7.37 and 7.32. Averages of all the products of the above with $w(k-1)$ are zero, because all terms are uncorrelated with $w(k-1)$, so we are now left with

$$a(k)E[x(k-1)\hat{x}(k-1)] = a[1-cb(k)]E[x(k-1)\hat{x}(k-1)]$$

giving

$$a(k) = a[1-cb(k)]$$

9 Calculation of $b(k)$ and $p(k)$ in section 7.3.2

To determine $b(k)$, we use equations 7.39 and 7.44. From equation 7.39 we have

$$p(k) = E[e^2(k)] = E\{e(k)[\hat{x}(k)-x(k)]\}$$

Substituting for $\hat{x}(k)$ from equation 7.38, and using the orthogonality equation 7.43 and 7.44 we obtain $p(k) = -E[e(k)x(k)]$.

From equations 7.44 and 7.37 we have

$$cE[e(k)x(k)] = -E[e(k)v(k)]$$

which enables the mean-square error to be written as

$$p(k) = \frac{1}{c}E[e(k)v(k)]$$

Substituting for $e(k) = \hat{x}(k)-x(k)$, and using equation 7.38, we obtain an expression for $p(k)$ containing three terms. Two of these, $E[\hat{x}(k-1)v(k)]$ and $E[x(k)v(k)]$, average to zero so we are left with

$$p(k) = \frac{1}{c}b(k)E[y(k)v(k)] = \frac{1}{c}\,b(k)\sigma_v^2$$

so $b(k) = c\,p(k)/\sigma_v^2$

To solve the problem completely we return to the mean-square error equation

$$\begin{aligned} p(k) &= E[\hat{x}(k)-x(k)]^2 \\ &= E\{a\hat{x}(k-1)+b(k)[y(k)-ac\hat{x}(k-1)]-x(k)\}^2 \end{aligned}$$

where we have used equation 7.46. Substituting for $y(k)$ from equation 7.37, and using equation 7.32, we can rewrite the above as

$$p(k) = E\{a[1-cb(k)]e(k-1)-[1-cb(k)]w(k-1)+b(k)v(k)\}^2$$

The cross products in the above expression average to zero because $e(k-1)$, $w(k-1)$ and $v(k)$ are independent of each other, so we have

$$p(k) = a^2[1-cb(k)]^2 p(k-1) + [1-cb(k)]^2 \sigma_w^2 + b^2(k)\sigma_v^2$$

where $p(k-1) = E[e^2(k-1)]$, $\sigma_w^2 = E[w^2(k-1)]$ and $\sigma_v^2 = E[v^2(k)]$. Substituting $p(k) = b(k)\sigma_v^2/c$ from the result derived earlier in this section of the appendix, we have

$$b(k)\{\sigma_v^2 + c^2[a^2 p(k-1) + \sigma_w^2]\} = c[a^2 p(k-1) + \sigma_w^2]$$

from which the solution, equation 7.47, is obtained. The other solution of the quadratic equation for $b(k)$ is $b(k) = 1/c$. This solution is neglected because it is time-invariant because c is constant, while the first solution is time-varying through $p(k-1)$.

10 FORTRAN program for Kalman filter in section 9.4

```
0           PROGRAM KALMAN
1           END
2                       DIMENSION P(2,2)
3                       DIMENSION Y(6), AK(2)
4                       X = 95.0
5                       V = 1.0
6                       P(1,1) = 10.0
7                       P(1,2) = 0.0
8                       P(2,1) = 0.0
9                       P(2,2) = 1.0
10                      Y(1) = 100.0
11                      Y(2) = 97.9
12                      Y(3) = 94.4
13                      Y(4) = 92.7
14                      Y(5) = 87.3
15                      Y(6) = 82.1
16                      DO 20 I = 1, 6
17                      X = X + V - 0.5
18                      V = V - 1.0
19                      P(1,1) = P(1,1) + P(2,2) + 2.0*P(2,1)
20                      P(2,1) = P(2,1) + P(2,2)
21                      D = P(1,1) + 1.0
22                      AK(1) = P(1,1)/D
23                      AK(2) = P(2,1)/D
24                      Z = Y(I) - X
25                      X = X + AK(1)*Z
26                      V = V + AK(2)*Z
27                      P(2,2) = P(2,2) - AK(2)*P(2,1)
28                      P(2,1) = P(2,1) - AK(2)*P(1,1)
29                      P(1,1) = (1.0 - AK(1))*P(1,1)
30                      WRITE (2,200) Y(I), X, V, P(1,1), P(2,1), P(2,2)
31                      WRITE (2,99)
32          20          CONTINUE
33          200         FORMAT (1X, 10(1H ), 7F15.4)
34          99          FORMAT (1X, 1H )
35                      STOP
36                      END
```

Comments

Steps 17, 18 are based on equation 9.19.
Steps 19, 20 come from equation 8.24 for $Q = 0$, matrix A as in equation 9.19, and denoting

$$P(1) = \begin{bmatrix} P(1,1) & P(1,2) \\ P(2,1) & P(2,2) \end{bmatrix}$$

where $P(1,2) = P(2,1)$ because of symmetry.
Steps 21, 22, 23 follow from the above and equation 8.23.

Steps 25, 26 are position and velocity estimates from equation 8.22.
Steps 27, 28, 29 come from equation 8.25 with $P_1(k)$ represented as above in steps 19, 20.

11 FORTRAN program for Wiener filter in section 9.5

```
0          PROGRAM WIENER
1          END
    C
    C      WIENER FILTER
    C
5                    DIMENSION H1(6,6),H2(6,6),H0(6),S(6),A(6,6),B(6,6),C(6,6)
6                    DIMENSION P1(6),P2(6),X1(6),X2(6),X(6),V(6),Y(6),X0(3),V0(3)
7                    SIG= 1.0
8                    X0(1) = 100.0
9                    X0(2) = 105.0
10                   X0(3) = 95.0
11                   V0(1) = 0.0
12                   V0(2) = 1.0
13                   V0(3) = -1.0
14                   DO 333 L = 1, 3
15                   DO 333 LP = 1, 3
16                   WRITE (2,99)
17                   WRITE (2,99)
18                   WRITE (2,731) X0(L), V0(LP)
19                   WRITE (2,99)
20                   DO 10 J = 1, 6
21                   X1(J) = X0(L) - 0.5*FLOAT(J)**2 + V0(LP)*FLOAT(J)
22         10        X2(J) = - FLOAT(J) + V0(LP)
    C
    C      CALCULATION OF THE MATRICES A, B, AND C
    C
26                   DO 12 I = 1, 6
27                   DO 11 J = 1, 6
28                   B(I,J) = X1(I)*X1(J) + FLOAT(I*J) + 10.0
29                   C(I,J) = X2(I)*X1(J) + FLOAT(J)
30         11        A(I,J) = B(I,J)
31         12        A(I,I) = A(I,I) + SIG
    C
    C      SOLUTION OF H1A = B AND H2A = C
    C
35                   DO 22 K = 1, 6
36                   DO 21 I = 1, 6
37         21        S(I) = B(K,I)
38                   CALL SOL(A, S, H0)
39                   DO 22 I = 1, 6
40         22        H1(K,I) = H0(I)
41                   DO 24 K = 1, 6
42                   DO 23 I = 1, 6
43         23        S(I) = C(K,I)
44                   CALL SOL(A, S, H0)
45                   DO 24 I = 1, 6
46         24        H2(K,I) = H0(I)
    C
    C      CALCULATION OF THE ESTIMATIONS
    C
50                   Y(1) = 100.0
51                   Y(2) = 97.9
52                   Y(3) = 94.4
53                   Y(4) = 92.7
54                   Y(5) = 87.3
55                   Y(6) = 82.1
56                   DO 30 K = 1, 6
57                   X(K) = 0.0
58                   V(K) = 0.0
59                   DO 30 I = 1, 6
60                   X(K) = X(K) + H1(K,I)*Y(I)
61         30        V(K) = V(K) + H2(K,I)*Y(I)
```

```
     C
     C     CALCULATION OF THE ERROR VARIANCES
     C
65                   DO 41 K = 1, 6
66                   P1(K) = X1(K)**2 + 10.0 +·FLOAT(K)**2
67                   P2(K) = X2(K)**2 + 1.0
68                   DO 41 I = 1, 6
69                   P1(K) = P1(K) - H1(K,I)*B(K,I)
70         41        P2(K) = P2(K) - H2(K,I)*C(K,I)
     C
     C     OUTPUTS
     C
74                   DO 50 K = 1, 6
75         50        WRITE (2,100) X(K), V(K), P1(K), P2(K)
76                   WRITE (2,99)
77         333       CONTINUE
78         99        FORMAT (1X, 1H )
79         731       FORMAT (1X, 3HX0=, F5.1, 6H  V0=, F4.1)
80         100       FORMAT (1X, 4F10.4)
81                   STOP
82                   END
     C
     C     END OF MAIN PROGRAM
     C
86                   SUBROUTINE SOL(A, Y, X)
     C
88                   DIMENSION A(6,6), X(6), Y(6)
89                   DO 502 IK = 1, 5
90                   K = 7 - IK
91                   K1 = K - 1
92                   DO 502 I = 1, K1
93                   Y(I) = Y(I) - Y(K)*A(K,I)/A(K,K)
94                   DO 502 J = 1, K1
95         502       A(J,I) = A(J,I) - A(J,K)*A(K,I)/A(K,K)
96                   X(1) = Y(1)/A(1,1)
97                   DO 503 I = 2, 6
98                   X(I) = Y(I)
99                   I1 = I - 1
100                  DO 504 J = 1, I1
101        504       X(I) = X(I) - X(J)*A(J,I)
102        503       X(I) = X(I)/A(I,I)
103                  RETURN
104                  END
     C
     C     END OF SUBROUTINE SOL
     C
```

Comments

Steps 21, 22 are based on equations 9.41 and 9.42.
Steps 28, 29 are based on equations 9.43 and 9.44.
Steps 30, 31 are based on equations 9.36, 9.37 and 9.43.
Steps 35 to 46 solve Wiener–Hopf equation for H1 and H2 using also the (recursive) subroutine in steps 86–104.
Steps 60, 61 represent the estimates from equations 9.28 and 9.29.
Steps 69, 70 represent the mean-square error terms given by equations 9.32 and 9.33.
Step 66 is a special case of equation 9.43 for $j = k$.
Step 67 represents equation 9.46.

References

1 Rabiner, L. R. *et al.* Terminology in digital signal processing. *IEEE Trans. Audio and Electroacoustics*, **AU-20**, 322–37, Dec. 1972
2 Steiglitz, K. *An introduction to discrete systems.* Wiley, 1974
3 Hovanessian, S. A. *et al. Digital computer methods in engineering.* McGraw-Hill, 1969
4 Cadzow, J. A. *Discrete-time systems; an introduction with interdisciplinary applications.* Prentice-Hall, 1973
5 Anderson, O. D. *Time series analysis and forecasting; the Box-Jenkins approach.* Butterworths, 1975
6 Freeman, H. *Discrete time systems.* Wiley, 1965
7 Jury, E. I. *Theory and applications of the* **z***-transform method.* Wiley, 1964
8 Robinson, E. A. *Statistical communication and detection.* Griffin, 1967
9 Kaiser, J. F. Digital filters. *Systems analysis by digital computer*, Kuo, F. F. and Kaiser, J. F. (editors). Wiley, 1966
10 Oppenheim, A. V. and Schaffer, R. W. *Digital signal processing.* Prentice-Hall, 1975
11 Stanley, W. D. *Digital signal processing.* Prentice-Hall, 1975
12 Gold, B. and Rader, C. M. *Digital processing of signals.* McGraw-Hill, 1969
13 Skolnik, M. I. *Introduction to radar systems.* McGraw-Hill, 1962
14 Helms, H. D. Nonrecursive digital filters. *IEEE Trans. on Audio and Electroacoustics*, **AU-16**, 3, 336–42, 1968
15 Rabiner, L. R. and Gold, B. *Theory and applications of digital signal processing.* Prentice-Hall, 1975
16 Linke, J. M. Residual attenuation equalization of broadband systems by generalized transversal networks. *Proc. IEE*, **114**, 3, 339–48, 1967
17 Bogner, R. E. and Constantinides, A. G. (editors) *Introduction to digital filtering.* Wiley, 1975
18 Haykin, S. S. A unified treatment of recursive digital filtering. *IEEE Trans. on Automatic Control*, 113–116, Feb. 1972
19 Smith, D. A. *et al.* Active bandpass filtering with bucket-brigade delay lines. *IEEE Journal of Solid-state circuits*, **SC-7**, 5, 421–5, Oct. 1972
20 Abramowitz, M. *et al.* (editors) *Handbook of Mathematical functions.* Dover Books, 1965
21 Schwartz, M. and Shaw, L. *Signal processing: Discrete spectral analysis detection and estimation.* McGraw-Hill, 1975
22 Mendel, J. M. *Discrete techniques of parameter estimation.* Dekker, 1973
23 Eykhoff, P. *System identification.* Wiley, 1974
24 Kalman, R. E. A new approach to linear filtering and prediction problems. *Trans. ASME, J. of Basic Engineering*, 35–45, March 1960
25 Sorenson, H. W. Least-squares estimation from Gauss to Kalman. *IEEE Spectrum*, **7**, 63–8, July 1970
26 Sorenson, H. W. Kalman filtering techniques. *Advances in control systems*, **3**, C. T. Leondes (editor). Academic Press, 1966
27 Sage, A. P. and Melsa, J. L. *Estimation theory with applications in communications and control.* McGraw-Hill, 1971
28 Scovell, G. *A guided tour through the implementations of a Kalman filter.* Univ. of Birmingham (England), Bosworth Course, April 1977
29 Jazwinski, A. H. *Stochastic processes and filtering theory.* Academic Press, 1970

30 Liebelt, P. B. *An introduction to optical estimation.* Addison-Wesley, 1967
31 Rousseliere, P. *Linear estimation theory methods.* M.Sc. project report, April 1978, Department of Electronic and Electrical Engineering, Univ. of Birmingham (England)
32 Singer, R. and Behnke, K. Real-Time Tracking Filter Evaluation and Selection for Tactical Applications. *IEEE Trans. Aerosp. Electron. Syst.*, **AES-7**, 100–110, Jan. 1971

Index

Accuracy considerations, 62
approximate inverse filter, 75
autoregressive moving-average filter, 28
autoregressive signal model, 100

Bilinear z-transformation
  alternative approach to, 65
  basic definition of, 51
bucket-brigade filter realization, 61

Computational steps in Kalman filter, 116
continuous- and discrete-time analysis, 3
convolution
  integral, 6
  summation, 6
covariance matrices, 114

Data vector equation, 112
delay cancellers in radar processing, 30
difference equations
  first-order, 4
  second-order, 5
  higher order, 16
digital filter
  canonic form, 24
  characteristics, 34
  classification, 27
  frequency response, 21
  realization schemes, 22
  serial and parallel connections, 26
  transfer function, 18
discrete Fourier series (DFS), 69
discrete Fourier transform (DFT), 71

Estimate
  of a constant, 126
  of random amplitude in noise, 125
estimator
  nonrecursive, 87
  optimum nonrecursive, 92
  optimum recursive, 100
  recursive, 89
  sample mean, 88

Finite impulse response (FIR) filter, 27
finite word length, 62
first-order RC filter, 4, 47, 53
Fourier series
  periodic frequency, 35
  periodic time, 35
Fourier transform
  discrete periodic series, 69
  finite time sequences, 71
frequency response of digital filter
  bandpass, 59, 60
  general, 21
  lowpass
    RC filter, 48, 54
    3-pole Butterworth, 50
    2-pole Butterworth, 56

Hamming window, generalized, 40
Hanning window, 40

Infinite impulse response (IIR) filter, 27
inverse filter
  concept of, 73
  optimum finite length, 75
inverse z-transformation, 19
inversion
  by long division, 19
  by partial-fraction expansion, 20
  by series expansion, 19

Kaiser window
  computer program, 147
  definition, 41
Kalman filter
  computer program for falling body problem, 151
  gain calculations, 128, 129
  scalar, 105
  vector, 115
  with driving force term, 132

Least mean-square, 77, 92, 102, 114

Matrix and vector dimensions, 116
minimum mean-square, 92, 102, 114
moving-average filter, 28

moving-target indicator, delay canceller, 30
multidimensional systems, 109

Nonrecursive estimator, 87
nonrecursive filter
  computer program, 147
  definition, 27, 83
  design procedure, 33, 36
  Fourier series application, 35
  linear phase types, 36
  transformations from lowpass to other types, 39
  unit-sample (impulse) response, 35
  with antisymmetrical impulse response, 36
  with symmetrical impulse response, 36
  with windows, 40

Optimum
  nonrecursive estimator, 92
  recursive estimator, 100
  recursive predictor, 105
orthogonality equations, 93, 103

Quantization errors, 63

Radar
  α–β tracker, 8
  canceller, 30
  tracking, 111, 112, 136–40
  transfer functions for α–β tracker, 31
  z-transform of α–β tracker, 14
recursive estimator, 89
  error, 149
  from the optimum nonrecursive estimator, 98
recursive filter
  basic definition, 27, 45
  bilinear z-transform, 52
  computer program, 148
  design considerations, 45
  first-order forms, 84
  frequency transformations, 57
  impulse invariance method, 46
  poles position design approach, 60
review of recursive and nonrecursive filters, 83
Ricatti difference equation, 116
round-off errors, 63

Sampled data frequency spectra, 13
sample mean estimator, 88
  error, 149
scalar Kalman
  filter, 105
  predictor, 108
scalar to matrix transformations, 114
second-order
  difference equation, 144
  resonator, 60
signal vector equation, 110
simultaneous filtering and prediction, 107, 120
state-space
  difference equation solution, 135
  notations, 125
s–z relationship, 12, 145

Table of transformations
  analogue to digital, 57
  scalar to matrix, 114

Unit pulse response, 17, 19, 35
unit step response, 18, 30

Vector Kalman
  filter, 115
  predictor, 119
vector Wiener filter, 123

Windows, 40, 41
  comparison in design, 43
Wiener filter
  computer program for falling body problem, 152
  disadvantages, 97
  scalar, 92
  vector, 123
Wiener–Hopf equation
  scalar, 94
  vector, 122

z-transform
  definition, 9
  final value, 15
  for convolution, 15
  for delayed sequence, 10
  Laplace definition of, 11
  relation to Laplace transform, 12
  table of commonly used sequences, 11